42nd Vietnam National Conference on Theoretical Physics (NCTP-42)

Journal of Physics: Conference Series Volume 1034

Can Tho, Vietnam
31 July – 3 August 2017

Editors:

Trinh Xuan Hoang
Hoang Ahn Tuan
Vu Ngoc Tuoc

ISBN: 978-1-5108-6409-2
ISSN: 1742-6588

Printed from e-media with permission by:

Curran Associates, Inc.
57 Morehouse Lane
Red Hook, NY 12571

Some format issues inherent in the e-media version may also appear in this print version.

Printed by Curran Associates, Inc. (2018)

For permission requests, please contact the Institute of Physics
at the address below.

Institute of Physics
Dirac House, Temple Back
Bristol BS1 6BE UK

Phone: 44 1 17 929 7481
Fax: 44 1 17 920 0979

techtracking@iop.org

Additional copies of this publication are available from:

Curran Associates, Inc.
57 Morehouse Lane
Red Hook, NY 12571 USA
Phone: 845-758-0400
Fax: 845-758-2633
Email: curran@proceedings.com
Web: www.proceedings.com

Table of contents

Volume 1034

42nd Vietnam National Conference on Theoretical Physics (NCTP-42)

31 July to 3 August 2017, Can Tho, Vietnam

Accepted papers received: 21 May 2018
Published online: 30 May 2018

Preface

Papers

IOP Conf. Series: Journal of Physics: Conf. Series **1034** (2018) 011001 doi:10.1088/1742-6596/1034/1/011001

Preface

The 42nd National Conference on Theoretical Physics (NCTP-42) was held during 31 July - 3 August 2017 in Can Tho, Vietnam.

The NCTP-42 was co-organized by the Institute of Physics, Vietnam Academy of Science and Technology (IOP-VAST) and Can Tho University (CTU) under the support of the Vietnamese Theoretical Physics Society (VTPS).

This meeting belongs to a series of annual theoretical physics conferences that started in 1976.

The conference has covered a wide range of theoretical physics topics from 4 major fields:

- Particle, nuclear and astro- physics,
- Molecular physics, quantum optics and quantum computation,
- Condensed matter physics,
- Soft matter, biological and interdisciplinary physics.

123 participants have participated in the conference. 2 invited talks, 24 oral and 74 poster contributions were presented.

This volume contains selected papers contributed by the participants.

Editors of the NCTP-42 Proceedings

Trinh Xuan Hoang, Hoang Anh Tuan and Vu Ngoc Tuoc

IOP Conf. Series: Journal of Physics: Conf. Series **1034** (2018) 011001 doi:10.1088/1742-6596/1034/1/011001

Organizers

- Institute of Physics, Vietnam Academy of Science and Technology (IOP-VAST)
- Can Tho University (CTU)

Sponsors

- Vietnam Academy of Science and Technology (VAST)

Honorary Chair

- Nguyen Van Hieu (VAST)

Chair

- Nguyen Ai Viet (IOP-VAST)

Organizing Committee

- Hoang Anh Tuan (IOP-VAST), Chair
- Bach Thanh Cong (VNU University of Science, Hanoi)
- Le Van Hoang (Ho Chi Minh City Pedagogical University)
- Dang Van Soa (Hanoi Metropolitan University)

Local Organizing Committee

- Nguyen Thanh Tien (Can Tho University), Chair
- Vu Thanh Tra (Can Tho University)
- Huynh Anh Huy (Can Tho University)

Program Committee

- Trinh Xuan Hoang (IOP-VAST), Chair
- Phung Van Dong (IOP-VAST)
- Ho Trung Dung (Ho Chi Minh city Institute of Physics)
- Nguyen The Toan (VNU University of Science, Hanoi)
- Vu Ngoc Tuoc (Hanoi University of Science and Technology)

Proceedings

- Trinh Xuan Hoang (IOP-VAST)
- Hoang Anh Tuan (IOP-VAST)
- Vu Ngoc Tuoc (Hanoi University of Science and Technology)

Secretariat

- Duong Thi Man (IOP-VAST)

IOP Conf. Series: Journal of Physics: Conf. Series **1034** (2018) 011001 doi:10.1088/1742-6596/1034/1/011001

List of Participants

Bach Thanh Cong
Faculty of Physics, VNU University of Science
congbachthanh@gmail.com

Bui Thi Ha Giang
Hanoi National University of Education
giang.giangsphn@gmail.com

Bui Dinh Hoi
University of Education, Hue University
buidinhhoi@gmail.com

Bui Phuong Thuy
Nam Dinh University of Technology Education
buiphuongthuyphys@gmail.com

Ca Nguyễn Anh Khoa
CanTho University
khoa.cn1984@gmail.com

Cao Huy Phuong
Hung Vuong University
caohuyphuongvlpt@yahoo.com.vn

Chu Thuy Anh
Institute of Physics, VAST
ctanh@iop.vast.ac.vn

Dang Van Soa
Hanoi Metropolitan University
soadangvan@gmail.com

Dao Thi Nhung
Institute For Interdisciplinary Research in Science and Education (IFIRSE)
dtnhung2008@gmail.com

Dao Vong Duc
Institute of Physics, VAST
dtman@iop.vast.ac.vn

Đinh Quang Vinh
Hanoi National university of Education
vinhhitech@gmail.com

Đỗ Quốc Tuấn
Faculty of Physics, Vietnam National University, Hanoi
tuanqdo@vnu.edu.vn

Đinh Thị Diệu Linh
Institute of Physics, VAST

IOP Conf. Series: Journal of Physics: Conf. Series **1034** (2018) 011001 doi:10.1088/1742-6596/1034/1/011001

ddlinh@iop.vast.ac.vn

Do Thi Huong
Institute of Physics, VAST
dthuong@iop.vast.ac.vn

Do Thi Hong Hai
Hanoi University of Mining and Geology
dohonghaimdc@gmail.com

Duong Thi Ha
Graduate University of Science and Technology, Vietnam Academy of Science and Technology
duongha@dhsptn.edu.vn

Duong Thi Nhu Tranh
HochiMinh City Univ. of Technology, Vietnam National University – Ho Chi Minh City
dtntranh@hcmut.edu.vn

Dương Thị Mân
Institute of Physics, VAST
duongthiman28@gmail.com

Hiền Đức Nguyễn
Sở GD&ĐT Gia Lai
n.duchien@gmail.com

Ho Sy Chuong
Dong Nai University
hosichuong@gmail.com

Hoang Anh - Tuan
Institute of Physics - VAST
hatuan@iop.vast.ac.vn

Hoang Do Ngoc Tram
HCMC University of Education
tramhdn@hcmup.edu.vn

Hoang Van Ngoc
Hanoi University of Science
conmai.bonghinh@gmail.com

Hoàng Hạnh Phương
Trường THPT Nam Sách,Hải Dương
trongnghia.hd@gmail.com

Hoàng Ngọc Long
Viện Vật Lý, VHLKHCNVN
hnlong@iop.vast.ac.vn

Hoàng Văn Hưng
Ho Chi Minh City University of Pedagogy
hunghv@hcmup.edu.vn

IOP Conf. Series: Journal of Physics: Conf. Series **1034** (2018) 011001 doi:10.1088/1742-6596/1034/1/011001

Huỳnh Vĩnh Phúc
Dong Thap University
huynhvinhphuc2710@gmail.com

Huynh Anh Huy
Can Tho University
hahuy@ctu.edu.vn

Huynh Thanh Duc
Ho Chi Minh City Institute of Physics, VAST
htduc@hcmip.vast.vn

Huynh Thi My Duyen
College of Natural Sciences, Can Tho University
huynhduyen1111@gmail.com

Lam Hoang Thai
Can Tho University
lamhoangthai.c3mtt@soctrang.edu.vn

Lê Đức Thiện
Viện Vật lý
leducthienhvt@gmail.com

Lê Hồng Việt
Tran Quoc Tuan University
levietmat@gmail.com

Lê Thị Ngọc Bảo
Trường Đại học Sư Phạm Huế
lethingocbao14@gmail.com

Lê Xuân Thùy
Vinh Long University of Technology Education
thuylx@vlute.edu.vn

Le Dinh
Hue University, College of Education
dinhle52@gmail.com

Le Duc Ninh
Institute For Interdisciplinary Research in Science and Education (IFIRSE)
leducninh@gmail.com

Le Tuan
Hanoi University of Science and Technology
letuan_2002@yahoo.com

Le Van-Hoang
Ho Chi Minh City University of Pedagogy
hoanglv2005@gmail.com

IOP Conf. Series: Journal of Physics: Conf. Series **1034** (2018) 011001 doi:10.1088/1742-6596/1034/1/011001

Le Duy Manh
Institute of Physics, VAST
ldmanh02468@gmail.com

Le Nguyen Tue Minh
Ho Chi Minh City University of Technology
tueminh11@gmail.com

Le Thi Cam Tu
Ho Chi Minh City University of Science
letu106@gmail.com

Le Thi Thu Phuong
University of Education, Hue University
thuphuonghueuni@gmail.com

Le Thi Hong Lien
Ha Noi University of Technology and Science
lien.lethihong@hust.edu.vn

Le Thi Quynh Huong
University of Khanh Hoa
lethiquynhhuong@yahoo.com

Le Tho Hue
Institute of Physics, VAST
lthue@iop.vast.ac.vn

Le Thu Thuy
Can Tho University
lethuthuy09a@gmail.com

Le Tri Dat
University of Science HCMC
tridat92@gmail.com

Le Trung Dung
University of Education, Hue University
letrungdung3112@gmail.com

Le Van Vinh
Hanoi University of Science and Technology
vinh.levan@hust.edu.vn

Le Van Xuan
Hanoi College of Technology and Economics
xuanlv.hn@gmail.com

Lương Văn Tùng
Đại học Đồng Tháp
lvtht1961@gmail.com

Mai Van Dung

IOP Conf. Series: Journal of Physics: Conf. Series **1034** (2018) 011001 doi:10.1088/1742-6596/1034/1/011001

Thu Dau Mot University
maivandung79tdm@gmail.com

Mẫn Văn Ngữ
Hung Yen Industrial Colleage
ngucdcnhy@gmail.com

Ngô Gia Vịnh
Sở Lao động và TBXH Bắc Ninh
ngogiavinh@bacninh.edu.vn

Ngô Thành Công
Ho Chi Minh City Institute of Physics
thanhcong2491@gmail.com

Ngo Son Tung
Ton Duc Thang University
ngosontung@tdt.edu.vn

Nguyễn Đăng Chiến
Dalat University
chiennd@dlu.edu.vn

Nguyễn Hoàng Giang
Comp. Phys. Lab, Ho Chi Minh City Univ. of Technology, Vietnam National University – Ho Chi Minh City
nhgiang@hcmut.edu.vn

Nguyễn Huy Việt
Viện Vật lý, Viện Hàn lâm Khoa học và Công nghệ Việt Nam
nhviet@iop.vast.ac.vn

Nguyễn Lâm Thùy Dương
Can Tho University
thuyduong01061991@gmail.com

Nguyễn Từ Niệm
VNU University of Science
nguyentuniem@gmail.com

Nguyễn Thị Hương
Đại học Thuỷ Lợi
nthuong@tlu.edu.vn

Nguyễn Văn Mện
An Giang University; University of Science Ho Chi Minh City
nguyenvanmenagu@gmail.com

Nguyen Mai Chung
Viện Vật Lý & Trường Đại học Khoa Học và Công Nghệ Hà Nội
maichung.n@gmail.com

Nguyen Ai Viet

IOP Conf. Series: Journal of Physics: Conf. Series **1034** (2018) 011001 doi:10.1088/1742-6596/1034/1/011001

Institute of Physics, VAST
vieta@iop.vast.ac.vn

Nguyen Anh Sang
Hanoi Pedagogical University No. 2
anhsangk13@gmail.com

Nguyen Dinh Hien
College of Education, Hue University
dinhhienphysics@gmail.com

Nguyen Hong Son
Trade Union University
atvsld@gmail.com

Nguyen Minh Hoa
Hue university of Medicine and Pharmacy
minhhoa2806@gmail.com

Nguyen Nhu Xuan
Học viện Kỹ thuật Quân sự
xuannn@mta.edu.vn

Nguyen Quang Bau
Faculty of Physics, Hanoi University of Science, VNU
nguyenquangbau54@gmail.com

Nguyen Thanh Nguyen
University of Science Ho Chi Minh City
nguyenthanhnguyen1196@gmail.com

Nguyen Thanh Tien
Can Tho University
thanhtienctu@gmail.com

Nguyen Thanh Tuong
University of Science, Ho Chi Minh City
nttuong.phys@gmail.com

Nguyen The Toan
Faculty of Physics, VNU University of Science, Vietnam National University
toannt@hus.edu.vn

Nguyen Thi Hoa
University of Transport and Communications
nthoaly@utc.edu.vn

Nguyen Thi Thao
Faculty of Physics, Hanoi National University of Education
ntthao.hnue@gmail.com

Nguyen Thi Xuan Huynh
University of Technology, VNU-HCM and Quy Nhon University

IOP Conf. Series: Journal of Physics: Conf. Series **1034** (2018) 011001 doi:10.1088/1742-6596/1034/1/011001

huynhqn@gmail.com

Nguyen Thi Bich Duyen
Institute of Physics
hoamai6hh@gmail.com

Nguyen Thi Thanh Ha
Hanoi University of Science and Technology
ha.nguyenthithanh1@hust.edu.vn

Nguyen Thi Thuy Nhung
Institute of Physics, Vietnam Academy of Science and Technology
ntnhung@iop.vast.vn

Nguyen Tri Lan
Institute of Physics, VAST
ntlan@iop.vast.ac.vn

Nguyen Truong Long
Can Tho University
truonglong@ctu.edu.vn

Nguyen Van Chuong
Le Quy Don Technical University
chuongnguyen11@gmail.com

Nguyen Van Hoa
Ho Chi Minh City University of Pedagogy
nvhoa62@yahoo.com

Nguyen Duy Vy
Ton Duc Thang University
nguyenduyvy@gmail.com

Ong Kim Le
University of Technology, VNU-HCM, Ho Chi Minh City, Vietnam
ongkimle1987@gmail.com

Pham Huu Kien
Thainguyen University of Education
phkien80@gmail.com

Pham Thi Bich Thao
Can Tho University
ptbthao@ctu.edu.vn

Pham Thi Thanh Nga
Thuyloi University
nga_ptt@tlu.edu.vn

Pham Tuan Vinh
Dong Thap University
tuanvinhdhdt@gmail.com

Pham Van Phuoc Em
Can Tho University
phuocemm0815007@gmailcom

Phan Thi Ngoc Loan
Ho Chi Minh City University of Education
loanphan@hcmup.edu.vn

Phan Van Cuong
Nha Trang University
cuongpv@ntu.edu.vn

Phù Chí Hòa
Dalat University
phchihoa@gmail.com

Phạm Khắc Hùng
Hanoi University of Science and Technology
thanhha.iep@gmail.com

Phung Van Dong
Institute of Physics, Vietnam Academy of Science and Technology
pvdong@iop.vast.ac.vn

Quach Kha Quang
Dong Thap University
quachkhaquang@gmail.com

Son Do
University of Technology, VNU-HCM
dnson@hcmut.edu.vn

Tiêu Tín Nguyên
Can Tho University
tieutinnguyen1100237@gmail.com

Tran Cong Phong
The Vietnam Institute of Educational Sciences
tcphong@moet.edu.vn

Tran Hai Hung
Nghe An college of education
haihung307@gmail.com

Tran Minh Tien
Thu Dau Mot University
tientm@tdmu.edu.vn

Tran Thanh Nhat
Graduate University of Science and Technology, Vietnam Academy of Science and Technology
tranthnhat@gmail.com

IOP Conf. Series: Journal of Physics: Conf. Series **1034** (2018) 011001 doi:10.1088/1742-6596/1034/1/011001

Tran Thi Nhan
Hanoi University of Industry
trannhan09@gmail.com

Tran Thi Thu Hanh
Hochiminh City University of Technology
thuhanhsp@gmail.com

Tran Van Quang
Department of Physics, University of Transport and Communications
tkuangv@gmail.com

Trần Đình Thám
Trường ĐH Phạm Văn Đồng, Quảng Ngãi
tdtham@pdu.edu.vn

Trần Thanh Dũng
Trường ĐH Thủ Dầu Một
ttdung24@gmail.com

Trần Thanh Tuyền
Trường THPT Quế Võ số 3
tranthanhtuyenc3qv3@bacninh.edu.vn

Trịnh Xuân Hoàng
Institute of Physics, VAST
hoang@iop.vast.ac.vn

Truong Minh Duc
Hue University, College of Education
tmduc2009@gmail.com

Truong Quoc Tuan
Dept. of Physics, Faculty of Natural Sci., Can Tho Univ.
truongquoctuanvl@gmail.com

Truong Xuan Nhut
Institute For Interdisciplinary Research in Science and Education (IFIRSE)
truongxuannhut@gmail.com

Vo Thanh Lam
Saigon University
vothanhlam2003@yahoo.com

Vo Trung Phuc
Can Tho University
votrungphuc063@gmail.com

Vo Van On
University of Thu Dau Mot
vovanon1963@gmail.com

Vu Ba Dung

Hanoi University of Minging and Geology
zung305@gmail.com

Vu Ngoc Tuoc
Hanoi Univ. of Science and Technology
tuoc.vungoc@hust.edu.vn

Vu Thanh Tra
Can Tho University
vttra@ctu.edu.vn

Vu Toan Thang
VNU University of Science, Hanoi
vuthang76@gmail.com

Zuoli Xiao
Peking University
zxiaojhu@163.com

IOP Conf. Series: Journal of Physics: Conf. Series **1034** (2018) 011002 doi:10.1088/1742-6596/1034/1/011002

Peer review statement

All papers published in this volume of *Journal of Physics: Conference Series* have been peer reviewed through processes administered by the proceedings Editors. Reviews were conducted by expert referees to the professional and scientific standards expected of a proceedings journal published by IOP Publishing.

IOP Conf. Series: Journal of Physics: Conf. Series **1034** (2018) 012001 doi:10.1088/1742-6596/1034/1/012001

On the validity of cosmic no-hair conjecture in an anisotropic inflationary model

Tuan Q Do

Faculty of Physics, VNU University of Science, Vietnam National University, Hanoi 120000, Vietnam

E-mail: tuanqdo@vnu.edu.vn

Abstract. We will present main results of our recent investigations on the validity of cosmic no-hair conjecture proposed by Hawking and his colleagues long time ago in the framework of an anisotropic inflationary model proposed by Kanno, Soda, and Watanabe. As a result, we will show that the cosmic no-hair conjecture seems to be generally violated in the Kanno-Soda-Watanabe model for both canonical and non-canonical scalar fields due to the existence of a non-trivial coupling term between scalar and electromagnetic fields. However, we will also show that the validity of the cosmic no-hair conjecture will be ensured once a unusual scalar field called the phantom field, whose kinetic energy term is negative definite, is introduced into the Kanno-Soda-Watanabe model.

1. Introduction

An inflationary universe has been regarded as one of leading paradigms of modern cosmology due to the fact that it has helped us to solve several important problems in cosmology such as the *monopole*, *horizon*, and *flatness* problems [1]. More interestingly, many theoretical predictions based on the inflation paradigm have been well confirmed by some high technology satellites such as the Wilkinson Microwave Anisotropy Probe (WMAP) [2] and Planck [3] built to observe the cosmic microwave background radiation (CMB). However, some anomalies detected in CMB such as the hemispherical asymmetry and the Cold Spot [2, 3] cannot be explained in the standard inflationary framework, which is based on the assumption that the spacetime of the early universe should be homogeneous and isotropic as the Friedmann-Lemaitre-Robertson-Walker (FLRW) metric. Therefore, some modifications to the metric or fields of standard inflationary models might be necessary. For example, the Bianchi metrics, which are homogeneous but anisotropic spacetimes [4], might be useful to realize the nature of the mentioned exotic features of CMB [5]. It is worth noting that some theoretical predictions of anisotropic inflationary models can be found in some earlier papers [6].

Note again that the early universe might be slightly anisotropic according to the recent observations of WMAP and Planck. Hence, it is natural to think about the state of the late-time universe. Will it be isotropic or still be slightly anisotropic is an open question to all of us. Theoretically, an important hint to this question can be obtained from the so-called cosmic no-hair conjecture proposed by Hawking and his colleagues long time ago [7, 8]. In particular, the statement of this conjecture is that our late-time universe would be simply homogeneous and isotropic, no matter the initial states and conditions of the early universe. However, a complete proof to this conjecture is not easy to achieve. Note that some partial proofs to this conjecture have been done in [9]. On the other hand, a counter-example to this conjecture has been claimed to exist in a supergravity motivated model by Kanno, Soda, and Watanabe (KSW) recently [10]. For recent interesting reviews on this model, see [11]. Hence, it is important to examine the validity of the cosmic no-hair conjecture in the KSW model. In particular, we should check if the conjecture is still violated when a canonical scalar field is replaced by non-canonical scalar ones such as the Dirac-Born-Infeld (DBI) [12], supersymmetric DBI (SDBI) [13], and covariant Galileon fields [14]. In addition, we should also check the fate of the conjecture when unusual field(s), e.g., the phantom field, one of alternative solutions to the dark energy problem in cosmology due to it negative kinetic term [15], is introduced into the KSW model. As a result, these scenarios have been investigated explicitly in our recent published papers [16, 17, 18, 19, 20].

The present article will be devoted to summarize basic results of our published papers. It will be organized as follows. A short introduction of our study has been written in the section 1. Non-canonical extensions of KSW model will be briefly shown in the section 2. The role of phantom field to the validity of cosmic no-hair conjecture will be discussed in the section 3. Finally, conclusions will be given in the section 4.

2. Non-canonical extensions of KSW model

2.1. Basic setup of KSW model

Before going to present details of the non-canonical extensions of KSW model mentioned above, we would like to draw here a very brief setup of KSW model [10, 11] since it would be very useful in order to compare the KSW model with its non-canonical extensions. As a result, an action of KSW model is given by [10, 11]

$$S_{\rm KSW} = \int d^4x\sqrt{-g}\left[\frac{M_p^2}{2}R - \frac{1}{2}\partial_\mu\phi\partial^\mu\phi - V\left(\phi\right) - \frac{1}{4}f^2\left(\phi\right)F_{\mu\nu}F^{\mu\nu}\right], \tag{1}$$

with M_p is the reduced Planck mass, ϕ is a canonical scalar field, and $F_{\mu\nu} \equiv \partial_\mu A_\nu - \partial_\nu A_\mu$ is the field strength of the electromagnetic (vector) field A_μ. In addition, $f(\phi)$ is a gauge kinetic function of $\phi \equiv \phi(t)$. Note that $f(\phi) = 1$ is always chosen in usual scenarios. As a result, varying this action with respect to $g^{\mu\nu}$ leads to the corresponding Einstein field equations:

$$M_p^2\left(R_{\mu\nu} - \frac{1}{2}Rg_{\mu\nu}\right) - \partial_\mu\phi\partial_\nu\phi + g_{\mu\nu}\left[\frac{1}{2}\partial^\sigma\phi\partial_\sigma\phi + V\left(\phi\right) + \frac{1}{4}f^2\left(\phi\right)F^{\rho\sigma}F_{\rho\sigma}\right] - f^2\left(\phi\right)F_{\mu\gamma}F_\nu^\gamma = 0. \tag{2}$$

Additionally, the field equations of vector and scalar fields read

$$\frac{\partial}{\partial x^\mu}\left[\sqrt{-g}f^2\left(\phi\right)F^{\mu\nu}\right] = 0, \tag{3}$$

$$\ddot{\phi} + 3H\dot{\phi} + \frac{\partial V\left(\phi\right)}{\partial\phi} + \frac{1}{2}f\left(\phi\right)\frac{\partial f\left(\phi\right)}{\partial\phi}F_{\mu\nu}F^{\mu\nu} = 0, \tag{4}$$

respectively. By taking the Bianchi type I metric,

$$ds^2 = -dt^2 + \exp\left[2\alpha\left(t\right) - 4\sigma\left(t\right)\right]dx^2 + \exp\left[2\alpha\left(t\right) + 2\sigma\left(t\right)\right]\left(dy^2 + dz^2\right), \tag{5}$$

along with the vector and scalar fields chosen as $A_\mu = (0, A_x\left(t\right), 0, 0)$ and $\phi = \phi\left(t\right)$, we are able to define the corresponding solution for the equation of vector field (3) as

$$\dot{A}_x\left(t\right) = f^{-2}\left(\phi\right)\exp\left[-\alpha - 4\sigma\right]p_A, \tag{6}$$

with p_A is a constant of integration. As a result, plugging this solution into the other field equations leads to

$$\ddot{\phi} = -3\dot{\alpha}\dot{\phi} - \frac{\partial V(\phi)}{\partial \phi} + f^{-3}(\phi)\frac{\partial f(\phi)}{\partial \phi}\exp\left[-4\alpha - 4\sigma\right]p_A^2, \quad (7)$$

$$\dot{\alpha}^2 = \dot{\sigma}^2 + \frac{1}{3M_p^2}\left[\frac{1}{2}\dot{\phi}^2 + V(\phi) + \frac{1}{2}f^{-2}(\phi)\exp\left[-4\alpha - 4\sigma\right]p_A^2\right], \quad (8)$$

$$\ddot{\alpha} = -3\dot{\alpha}^2 + \frac{1}{M_p^2}V(\phi) + \frac{1}{6M_p^2}f^{-2}(\phi)\exp\left[-4\alpha - 4\sigma\right]p_A^2, \quad (9)$$

$$\ddot{\sigma} = -3\dot{\alpha}\dot{\sigma} + \frac{1}{3M_p^2}f^{-2}(\phi)\exp\left[-4\alpha - 4\sigma\right]p_A^2, \quad (10)$$

Now, by choosing the exponential potential and gauge kinetic function,

$$V(\phi) = V_0 \exp\left[\frac{\lambda}{M_p}\phi\right];\ f(\phi) = f_0 \exp\left[\frac{\rho}{M_p}\phi\right], \quad (11)$$

along with the following ansatz,

$$\alpha = \zeta \log(t);\ \sigma = \eta \log(t);\ \frac{\phi}{M_p} = \xi \log(t) + \phi_0, \quad (12)$$

ones have been able to define the corresponding solutions for scale factors of metric from the above field equations such as [10]

$$\zeta = \frac{\lambda^2 + 8\rho\lambda + 12\rho^2 + 8}{6\lambda(\lambda + 2\rho)};\ \eta = \frac{\lambda^2 + 2\rho\lambda - 4}{3\lambda(\lambda + 2\rho)}. \quad (13)$$

Here λ, ρ, V_0, f_0, and ϕ_0 are positive constants. It is noted that σ acts as a deviation from the isotropy, i.e., $\sigma \ll \alpha$ in order to be consistent with the data of WMAP and Planck. Consequently, $\eta \ll \zeta$ is required for the obtained anisotropic inflationary solution. As a result, this constraint will be easily fulfilled if $\rho \gg \lambda \sim \mathcal{O}(1)$. By converting the field equations into autonomous equations of dynamical variables defined as

$$X = \frac{\dot{\sigma}}{\dot{\alpha}};\ Y = \frac{\dot{\phi}}{M_p\dot{\alpha}};\ Z = \frac{1}{f_0 M_p \dot{\alpha}}\exp\left[-\frac{\rho}{M_p}\phi - 2\alpha - 2\sigma\right]p_A, \quad (14)$$

ones have shown that the anisotropic power-law solution found above is indeed equivalent to an anisotropic fixed point of dynamical system, which is a non-trivial solution of $dX/d\alpha = dY/d\alpha = dZ/d\alpha = 0$,

$$X = \frac{2\left(\lambda^2 + 2\rho\lambda - 4\right)}{\lambda^2 + 8\rho\lambda + 12\rho^2 + 8};\ Y = -\frac{12\left(\lambda + 2\rho\right)}{\lambda^2 + 8\rho\lambda + 12\rho^2 + 8},$$
$$Z^2 = \frac{18\left(\lambda^2 + 2\rho\lambda - 4\right)\left(-\lambda^2 + 4\rho\lambda + 12\rho^2 + 8\right)}{\left(\lambda^2 + 8\rho\lambda + 12\rho^2 + 8\right)^2}. \quad (15)$$

Moreover, ones have shown that this anisotropic fixed point, or the corresponding anisotropic power-law solution, is indeed a stable and attractor solution during the inflationary phase [10]. This implies that the late-time universe in the KSW model will be anisotropic rather than isotropic, meaning that the cosmic no-hair conjecture is violated in the KSW model. Interested readers are encouraged to read papers in [10] for more technical details. Additionally, cosmological implications of KSW model, e.g., imprints of anisotropic inflation on the CMB through correlations between T, E, and B modes, can be found in the recent interesting review papers in [11]. In the following subsections, we will present the non-canonical extensions of KSW model mentioned in the introduction section to see whether the cosmic no-hair conjecture holds.

IOP Conf. Series: Journal of Physics: Conf. Series **1034** (2018) 012001 doi:10.1088/1742-6596/1034/1/012001

2.2. DBI model

In this subsection, we will present basic results of an interesting extension of KSW model, the DBI model [12], in which the canonical scalar field ϕ is replaced by the non-canonical DBI one [17] as follows

$$S_{\rm DBI} = \int d^4x\sqrt{-g}\left[\frac{M_p^2}{2}R + \frac{1}{f(\phi)}\frac{\gamma-1}{\gamma} - V(\phi) - \frac{1}{4}h^2(\phi)F_{\mu\nu}F^{\mu\nu}\right], \tag{16}$$

with $\gamma \equiv 1/\sqrt{1+f(\phi)\partial_\mu\phi\partial^\mu\phi}$ is the Lorentz factor characterizing the motion of the D brane [12]. It is clear that $S_{\rm DBI}$ will reduce to $S_{\rm KSW}$ once $\gamma \to 1$ (or equivalently $f(\phi) \to 0$). It appears that $\gamma \geq 1$ for non-negative $f(\phi)$ if $\phi = \phi(t)$. Note that $h(\phi)$ in this model is identical to $f(\phi)$ in the KSW model. As a result, the following field equations turn out to be [17]

$$\ddot{\phi} = -\frac{3\dot{\alpha}}{\gamma^2}\dot{\phi} - \frac{\partial_\phi V}{\gamma^3} - \frac{\partial_\phi f}{2f}\frac{(\gamma+2)(\gamma-1)}{(\gamma+1)\gamma}\dot{\phi}^2 + \frac{h^{-3}\partial_\phi h}{\gamma^3}\exp[-4\alpha-4\sigma]p_A^2, \tag{17}$$

$$\dot{\alpha}^2 = \dot{\sigma}^2 + \frac{1}{3}\left[\frac{\gamma^2}{\gamma+1}\dot{\phi}^2 + V(\phi) + \frac{1}{2}h^{-2}\exp[-4\alpha-4\sigma]p_A^2\right], \tag{18}$$

$$\ddot{\alpha} = -3\dot{\alpha}^2 + \left[\frac{\gamma(\gamma-1)}{2(\gamma+1)}\dot{\phi}^2 + V(\phi)\right] + \frac{1}{6}h^{-2}\exp[-4\alpha-4\sigma]p_A^2, \tag{19}$$

$$\ddot{\sigma} = -3\dot{\alpha}\dot{\sigma} + \frac{1}{3}h^{-2}\exp[-4\alpha-4\sigma]p_A^2, \tag{20}$$

here we have used the same setup for fields and metric as proposed in the KSW model. As a result, the corresponding power-law anisotropic solution turns out to be [17]

$$\zeta = \frac{\lambda^2+8\rho\lambda+12\rho^2+8\gamma_0}{6\lambda(\lambda+2\rho)};\ \eta = \frac{\lambda^2+2\rho\lambda-4\gamma_0}{3\lambda(\lambda+2\rho)}. \tag{21}$$

which is equivalent with an anisotropic fixed point of dynamical system defined as

$$X = \frac{2[\hat{\gamma}_0\lambda(\lambda+2\rho)-4]}{\hat{\gamma}_0(\lambda^2+8\lambda\rho+12\rho^2)+8};\ Y = -\frac{12\hat{\gamma}_0(\lambda+2\rho)}{\hat{\gamma}_0(\lambda^2+8\lambda\rho+12\rho^2)+8},$$

$$Z^2 = \frac{18[\hat{\gamma}_0\lambda(\lambda+2\rho)-4][\hat{\gamma}_0(-\lambda^2+4\lambda\rho+12\rho^2)+8]}{[\hat{\gamma}_0(\lambda^2+8\lambda\rho+12\rho^2)+8]^2};\ \hat{\gamma}_0 = \gamma_0^{-1}, \tag{22}$$

here γ has been fixed to be a constant, i.e., $\gamma = \gamma_0 \geq 1$. It is clear that these solutions will recover that of KSW model shown above in the limit $\gamma_0 \to 1$. Of course, $\rho \gg \lambda \sim \mathcal{O}(1)$ is also required to have the anisotropic inflationary solution, similar to the KSW model. And using the dynamical system approach, we have been able to show that the anisotropic power-law solution of DBI model is also stable and attractive as that of KSW model during the inflationary phase [17] (see the plot listed below for more details). Hence, we can conclude that the cosmic no-hair conjecture is also violated in the DBI extension of KSW model.

2.3. SDBI model

As a result, an action of SDBI extension of KSW model is given by [13, 18]

$$S_{\rm SDBI} = \int d^4x\sqrt{-g}\left[\frac{M_p^2}{2}R + \frac{1}{f(\phi)}\frac{\gamma-1}{\gamma} - \Sigma_0^2\,U(\phi) - \frac{1}{4}h^2(\phi)F_{\mu\nu}F^{\mu\nu}\right], \tag{23}$$

with

$$\Sigma_0(\gamma) = \left(\frac{\gamma+1}{2\gamma}\right)^{1/3} \leq 1;\ U(\phi) = \left(\frac{27}{2f(\phi)}\right)^{\frac{1}{3}} \left(\frac{dW(\phi)}{d\phi}\right)^{\frac{4}{3}}. \tag{24}$$

Here $W(\phi)$ plays as a super-potential. For a detailed derivation of this action, one can see [13]. As a result, the corresponding field equations of SDBI model are given by [18]

$$\begin{aligned}
\Sigma_1\ddot{\phi} &= -3\Sigma_2\dot{\alpha}\dot{\phi} - \left(\Sigma_0^2 + \frac{\gamma}{3\Sigma_0} f\dot{\phi}^2\right)\partial_\phi U - \frac{\Sigma_1 - \gamma^3}{2}\frac{\partial_\phi f}{f}\dot{\phi}^2 - \frac{\partial_\phi f}{2f}\frac{(\gamma+2)(\gamma-1)\gamma^2}{\gamma+1}\dot{\phi}^2 \\
&\quad + h^{-3}\partial_\phi h \exp[-4\alpha - 4\sigma]\, p_A^2, &(25)\\
\dot{\alpha}^2 &= \dot{\sigma}^2 + \frac{1}{3}\left[\frac{\gamma^2}{\gamma+1}\dot{\phi}^2 + \frac{2\gamma+1}{3}\Sigma_0^2 U + \frac{h^{-2}}{2}\exp[-4\alpha - 4\sigma]\, p_A^2\right], &(26)\\
\ddot{\alpha} &= -3\dot{\alpha}^2 + \frac{\gamma(\gamma-1)}{2(\gamma+1)}\dot{\phi}^2 + \frac{\gamma+2}{3}\Sigma_0^2 U + \frac{h^{-2}}{6}\exp[-4\alpha - 4\sigma]\, p_A^2, &(27)\\
\ddot{\sigma} &= -3\dot{\alpha}\dot{\sigma} + \frac{h^{-2}}{3}\exp[-4\alpha - 4\sigma]\, p_A^2. &(28)
\end{aligned}$$

with

$$\Sigma_1 = \gamma^3 + \frac{\gamma}{9\Sigma_0}\left(3\gamma^2 + \gamma - 1\right) fU;\ \Sigma_2 = \gamma + \frac{\gamma}{3\Sigma_0} fU. \tag{29}$$

Hence, the following Bianchi type I power-law solution of this model turns out to be [18]

$$\begin{aligned}
\zeta &= \frac{N - \sqrt{N^2 - 4MP}}{2M};\ \eta = -\zeta + \frac{\rho}{\lambda} + \frac{1}{2}, &(30)\\
M &= 18\lambda^2\left(\gamma_0^2 - 1\right);\ N = 3\lambda(\gamma_0 + 1)\left[(5\gamma_0 + 1)\lambda + 6(\gamma_0 + 1)\rho\right], &(31)\\
P &= (\gamma_0 + 1)\left[(2\gamma_0 + 1)\lambda^2 + 2(5\gamma_0 + 7)\lambda\rho + 12(\gamma_0 + 2)\rho^2\right] + 8\gamma_0(5\gamma_0 + 1). &(32)
\end{aligned}$$

Similar to the DBI model, we are able to define the following anisotropic fixed point of SDBI model to be

$$X = \frac{-S_2 - \sqrt{S_2^2 + 16S_1S_3}}{4S_1};\ Y = -\frac{4}{\lambda + 2\rho}(X+1);\ Z^2 = 3X(1-2X) - 3\rho XY, \tag{33}$$

with

$$\begin{aligned}
S_1 &= (\hat{\gamma}_0 + 1)\left[(\hat{\gamma}_0 + 2)\lambda^2 + 2(7\hat{\gamma}_0 + 5)\lambda\rho + 12(2\hat{\gamma}_0 + 1)\rho^2\right] + 8(\hat{\gamma}_0 + 5), &(34)\\
S_2 &= (\hat{\gamma}_0 + 1)\left[(\hat{\gamma}_0 - 7)\lambda^2 + 8(4\hat{\gamma}_0 - 1)\lambda\rho + 12(5\hat{\gamma}_0 + 1)\rho^2\right] + 32(\hat{\gamma}_0 + 5), &(35)\\
S_3 &= (\hat{\gamma}_0 + 1)\left[(5\hat{\gamma}_0 + 1)\lambda^2 + 4(4\hat{\gamma}_0 - 1)\lambda\rho + 12(\hat{\gamma}_0 - 1)\rho^2\right] - 8(\hat{\gamma}_0 + 5). &(36)
\end{aligned}$$

Here $\gamma = \gamma_0 =$ constant and $\hat{\gamma}_0 = \gamma_0^{-1}$. It is noted that γ_0 in the SDBI model has been required to obey the following constraint [18]

$$1 \leq \gamma_0 < 1 + \frac{\lambda}{\rho}. \tag{37}$$

in order to have the anisotropic power-law solution. This result is consistent with the investigation for isotropic solutions done in [13]. Note again that γ_0 can be arbitrarily larger than one in the DBI extension of KSW model [17]. In addition, γ in the DBI model is expected to be larger than one in order to have the significant primordial non-Gaussianity in CMB [12]. However, the observations of WMAP and Planck have not confirmed the existence of the CMB non-Gaussianity. Hence, $\gamma \simeq 1$ is cosmologically reasonable. For the stability of the obtained anisotropic power-law inflationary solution, we have used the dynamical system approach to investigate and arrived at a conclusion that it is stable and attractor solution, similar to that found in the KSW and DBI models [18]. For instance, one can see the plot listed below for numerical results showing the attractor behavior of the obtained solution.

2.4. Galileon model

Next, we will consider another interesting extension of KSW model [20], in which both kinetic and potential terms of the canonical scalar field are replaced by non-canonical terms of the so-called covariant Galileon model [14] as follows

$$S_{\rm G} = \int d^4x\sqrt{-g}\left\{\frac{M_p^2}{2}R + k_0\exp\left[\frac{\tau\phi}{M_p}\right]X - g_0\exp\left[\frac{\lambda\phi}{M_p}\right]X\Box\phi - \frac{f_0^2}{4}\exp\left[-\frac{2\rho\phi}{M_p}\right]F_{\mu\nu}F^{\mu\nu}\right\}, \tag{38}$$

here $X \equiv -\frac{1}{2}\partial^\mu\phi\partial_\mu\phi$. As a result, the corresponding field equations of this model turn out to be [14, 20]

$$\mathcal{E}^{(2)} + \mathcal{E}^{(3)} + f^{-3}f_\phi\exp\left[-4\alpha - 4\sigma\right]p_A^2 = 0, \tag{39}$$

$$-3M_p^2\left(\dot\alpha^2 - \dot\sigma^2\right) + \frac{k_0}{2}\exp\left[\frac{\tau\phi}{M_p}\right]\dot\phi^2 + 3g_0\exp\left[\frac{\lambda\phi}{M_p}\right]\dot\alpha\dot\phi^3 - \frac{g_0\lambda}{2M_p}\exp\left[\frac{\lambda\phi}{M_p}\right]\dot\phi^4$$
$$+\frac{f^{-2}}{2}\exp\left[-4\alpha - 4\sigma\right]p_A^2 = 0, \tag{40}$$

$$-M_p^2\ddot\alpha - 3M_p^2\dot\alpha^2 + \frac{g_0}{2}\exp\left[\frac{\lambda\phi}{M_p}\right]\left(\ddot\phi + 3\dot\alpha\dot\phi\right)\dot\phi^2 + \frac{f^{-2}}{6}\exp\left[-4\alpha - 4\sigma\right]p_A^2 = 0, \tag{41}$$

$$\ddot\sigma + 3\dot\alpha\dot\sigma - \frac{f^{-2}}{3M_p^2}\exp\left[-4\alpha - 4\sigma\right]p_A^2 = 0. \tag{42}$$

where

$$\mathcal{E}^{(2)} = k_0\left(\Box\phi - \frac{\tau}{M_p}X\right)\exp\left[\frac{\tau\phi}{M_p}\right], \tag{43}$$

$$\mathcal{E}^{(3)} = g_0\left\{\frac{\lambda^2}{M_p^2}\frac{\dot\phi^2}{2} - \left[6H\frac{\ddot\phi}{\dot\phi} + 9H^2 - \left(\sum_{i=1}^{3}H_i^2\right) - R_{00}\right] + \frac{2\lambda}{M_p}\ddot\phi\right\}\dot\phi^2\exp\left[\frac{\lambda\phi}{M_p}\right]. \tag{44}$$

It is noted that $H \equiv (H_1+H_2+H_3)/3$ is the mean Hubble parameter along with its components $H_i \equiv \dot a_i/a_i$ $(i = 1-3)$. As a result, the corresponding solutions of this model can be defined to be [20]

$$\zeta = \frac{5}{12} + \frac{z}{2} + \frac{\sqrt{\Delta}}{12};\ \eta = -\zeta + z + \frac{1}{2}, \tag{45}$$

with

$$z \equiv \frac{\rho}{\lambda};\ \Delta \equiv 9 - \frac{64k_0}{\lambda^2} - 60z^2 - 20z, \tag{46}$$

provided that $k_0 \leq -3\lambda^2z^2/4$. Similar to the KSW model and its (S)DBI extensions, ρ must be much larger than λ to ensure the existence of inflationary solutions [20]. As a result, if $\rho \gg \lambda$ we will have the following approximations such as

$$\zeta \simeq z \gg 1;\ \eta \simeq \frac{1}{12};\ k_0 \simeq -\frac{3\rho^2}{2}. \tag{47}$$

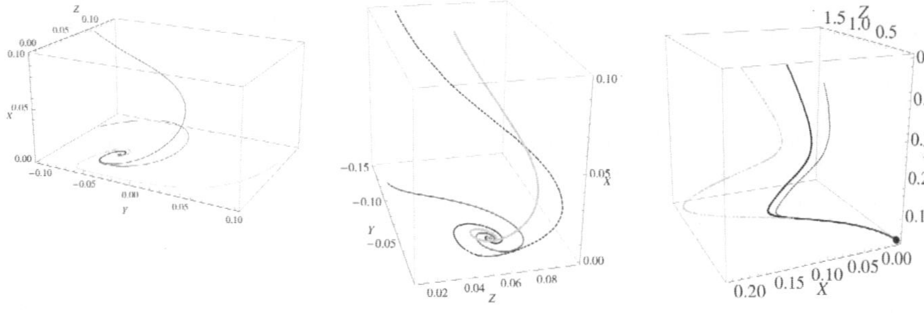

Figure 1. (From left to right) Attractor behavior of the anisotropic fixed points of the DBI, SDBI, and Galileon models, respectively. Here, the field parameters are chosen as $\lambda = 0.1$, $\rho = 50$, and $\gamma = 1.5$ for the DBI model; $\lambda = 0.1$, $\rho = 50$, and $\gamma = 1.0001$ for the SDBI model; $\lambda = 0.1$, $\rho = 50$, and $k_0 = -3\rho^2/2$ for the Galileon model. Note that these pictures are taken from papers [17], [18], and [20], respectively.

Additionally, the corresponding anisotropic fixed point of this model turns out to be

$$X = \frac{-12z^2 + 8z + 7 - 32\frac{k_0}{\lambda^2} - 3\left(2z+1\right)\sqrt{\Delta}}{8\left(6z^2 + 5z + 1 + 4\frac{k_0}{\lambda^2}\right)}; \; Y = \frac{4\left(X+1\right)}{\lambda + 2\rho}; \; Z^2 = -3X\left(2X - \rho Y - 1\right). \tag{48}$$

Furthermore, the stability analysis and numerical result have confirmed that the anisotropic power-law inflationary solution of Galileon model is indeed stable and attractive as expected. Indeed, one can see the Fig. 1 for the numerical confirmation.

3. The role of phantom field to the validity of cosmic no-hair conjecture

We have shown above that the cosmic no-hair conjecture seems to be generally violated in the framework of KSW model, even when the scalar field is of non-canonical forms such as the DBI, SDBI, and Galileon fields, due to the existence of the unusual coupling term between scalar and vector fields, $f^2(\phi)F^2$. Hence, the validity of the cosmic no-hair conjecture in unusual scenarios might need the existence of unusual fields. Indeed, by introducing a phantom (scalar) field, whose kinetic energy is negative definite [15], we have been able to show that the corresponding anisotropic power-law inflationary solutions are indeed unstable rather than stable [16]. This means that the validity of the cosmic no-hair conjecture really needs the existence of the phantom field in the KSW model [16, 19] and its non-canonical extensions [17, 18, 20]. Now, we would like to briefly demonstrate here how unstable modes appear once the phantom field is introduced. Let us begin with an action of two-scalar-field extension of KSW model [16]

$$S = \int d^4x\sqrt{-g}\left[\frac{M_p^2}{2}R - \frac{1}{2}\partial_\mu\phi\partial^\mu\phi + \frac{1}{2}\partial_\mu\psi\partial^\mu\psi - V_1(\phi) - V_2(\psi) - \frac{1}{4}f_1^2(\phi)f_2^2(\psi)F_{\mu\nu}F^{\mu\nu}\right]. \tag{49}$$

As a result, the following field equations are given by

$$\ddot{\phi} = -3\dot{\alpha}\dot{\phi} - \frac{\partial V_1}{\partial\phi} + f_1^{-3}f_2^{-2}\frac{\partial f_1}{\partial\phi}\exp\left[-4\alpha - 4\sigma\right]p_A^2, \tag{50}$$

$$\ddot{\psi} = -3\dot{\alpha}\dot{\psi} + \frac{\partial V_2}{\partial\psi} - f_1^{-2}f_2^{-3}\frac{\partial f_2}{\partial\psi}\exp\left[-4\alpha - 4\sigma\right]p_A^2, \tag{51}$$

$$\dot{\alpha}^2 = \dot{\sigma}^2 + \frac{1}{3M_p^2}\left[\frac{1}{2}\dot{\phi}^2 - \frac{1}{2}\dot{\psi}^2 + V_1 + V_2 + \frac{f_1^{-2}f_2^{-2}}{2}\exp\left[-4\alpha - 4\sigma\right]p_A^2\right], \tag{52}$$

$$\ddot{\alpha} = -3\dot{\alpha}^2 + \frac{1}{M_p^2}(V_1 + V_2) + \frac{f_1^{-2}f_2^{-2}}{6M_p^2}\exp\left[-4\alpha - 4\sigma\right]p_A^2, \tag{53}$$

$$\ddot{\sigma} = -3\dot{\alpha}\dot{\sigma} + \frac{f_1^{-2}f_2^{-2}}{3M_p^2}\exp\left[-4\alpha - 4\sigma\right]p_A^2. \tag{54}$$

And by choosing exponential forms of potentials and gauge kinetic functions such as [16]

$$V_1(\phi) = V_{01}\exp\left[\frac{\lambda_1\phi}{M_p}\right];\ V_2(\psi) = V_{02}\exp\left[\frac{\lambda_2\psi}{M_p}\right],$$
$$f_1(\phi)f_2(\psi) = f_0\exp\left[\frac{\rho_1\phi}{M_p} + \frac{\rho_2\psi}{M_p}\right], \tag{55}$$

along with the following ansatz,

$$\alpha = \zeta\log(t);\ \sigma = \eta\log(t);\ \frac{\phi}{M_p} = \xi_1\log(t) + \phi_0;\ \frac{\psi}{M_p} = \xi_2\log(t) + \psi_0, \tag{56}$$

we have been able to figure out a set of anisotropic power-law solutions as [16]

$$\zeta = \frac{4(\lambda_1\rho_2 + \lambda_2\rho_1)(2\lambda_1\lambda_2 + 3\lambda_1\rho_2 + 3\lambda_2\rho_1) + \lambda_1^2\lambda_2^2 + 8(\lambda_2^2 - \lambda_1^2)}{6\lambda_1\lambda_2(\lambda_1\lambda_2 + 2\lambda_1\rho_2 + 2\lambda_2\rho_1)},$$
$$\eta = \frac{\lambda_1\lambda_2(\lambda_1\lambda_2 + 2\lambda_1\rho_2 + 2\lambda_2\rho_1) - 4(\lambda_2^2 - \lambda_1^2)}{3\lambda_1\lambda_2(\lambda_1\lambda_2 + 2\lambda_1\rho_2 + 2\lambda_2\rho_1)}.$$

And the corresponding inflationary solutions will approximately be

$$\zeta \simeq \frac{\rho_1}{\lambda_1} + \frac{\rho_2}{\lambda_2} \gg 1;\ \eta \simeq \frac{1}{3}, \tag{57}$$

provided that $\rho_i \gg \lambda_i \sim \mathcal{O}(1)$. Next, we use the power-law perturbations [16],

$$\delta\alpha,\ \delta\sigma,\ \delta\phi,\ \delta\psi \propto t^n, \tag{58}$$

which are compatible with the power-law solutions to study the stability of the obtained solutions. It is noted that $n > 0$ and $n < 0$ will correspond to unstable and stable perturbation mode(s), respectively. As a result, the following of equation of n turns out to be [16]

$$f(n) \equiv n^7 + b_7n^6 + b_6n^5 + b_5n^4 + b_4n^3 + b_3n^2 + b_2n + b_1 = 0, \tag{59}$$

with

$$b_1 = -\frac{2vl}{\lambda_1}\left\{\left[\lambda_1^2\lambda_2^2(5\zeta - \eta - 1) + 2\lambda_1\lambda_2(\lambda_1\rho_2 + \lambda_2\rho_1)(3\zeta - 3\eta - 1)\right.\right.$$
$$\left.\left. +4\left(\lambda_1^2 - \lambda_2^2\right)\right]\lambda_1 u + 8\lambda_2\rho_1\rho_2(3\lambda_1\rho_1\zeta - 3\lambda_1\rho_1\eta - \lambda_1\rho_1 - 2)\,l\right\}, \tag{60}$$

where

$$u = \frac{V_{01}}{M_p^2}\exp[\lambda_1\phi_0];\ v = \frac{V_{02}}{M_p^2}\exp[\lambda_2\psi_0];\ l = \frac{p_A^2}{M_p^2f_0^2}\exp[-2(\rho_1\phi_0 + \rho_2\psi_0)]. \tag{61}$$

It is straightforward to see that $b_1 < 0$ during the inflationary phase with $\zeta \gg \eta$, $u > 0$, $v > 0$, and $l > 0$. Furthermore, we have observed in [16] that $f(n = 0) = b_1 < 0$, while $f(n) \sim n^7 > 0$ as $n \gg 1$. This indicates that the curve $f(n)$ will cross the positive horizontal axis at least one time, i.e., the equation $f(n) = 0$ will admit at least one positive root $n > 0$ [16]. Hence, the corresponding anisotropic power-law solution is really unstable during the inflationary phase as expected. Note that the same conclusion can be found in the other non-canonical extensions of KSW model, i.e., the DBI, SDBI, and Galileon models [17, 18, 20], as well as in an extension of two-scalar-field model, in which a mixed kinetic term of canonical and phantom fields, $\partial_\mu \phi \partial^\mu \psi$, is involved [19].

4. Conclusions

We have presented basic details of our recent studies on the fate of the cosmic no-hair conjecture in the context of KSW model. In particular, we have shown that the KSW model always admits the Bianchi type I metric as its power-law stable and attractor solutions for both canonical and non-canonical scalar fields due to the existence of the unusual coupling term between the scalar and vector fields. This indicates that the Hawking cosmic no-hair conjecture seems to be generally violated in the context of KSW model. However, we have also shown that the conjecture might be valid if the phantom field with its negative kinetic term is involved into the KSW model. As a result, the existence of phantom field will introduce at least one unstable mode to field perturbations, and therefore any initial spatial anisotropy of Bianchi type I metric will decay as the cosmic no-hair conjecture states. Note that this conclusion has been shown to be valid for all non-canonical extensions of KSW model mentioned above [17, 18, 19, 20]. Hence, we come to a conclusion that the phantom field might play an important role in order to ensure the validity of the cosmic no-hair conjecture.

Acknowledgments

The author is deeply grateful to Professor W. F. Kao of Institute of Physics in National Chiao Tung University for his useful advice on the KSW anisotropic inflation model and the cosmic no-hair conjecture. The author would like to thank Dr. Ing-Chen Lin as well as Dr. Sonnet Hung Q. Nguyen very much for their collaboration. This research is supported in part by the Vietnam National Foundation for Science and Technology Development (NAFOSTED) under Grant No. 103.01-2017.12.

References

[1] Guth A H 1981 The inflationary universe: A possible solution to the horizon and flatness problems *Phys. Rev.* D **23** 347
Linde A D 1982 A new inflationary universe scenario: A possible solution of the horizon, flatness, homogeneity, isotropy and primordial monopole problems *Phys. Lett.* B **108** 389
Linde A D 1983 Chaotic inflation *Phys. Lett.* B **129** 177
[2] Komatsu E et al. [WMAP Collaboration] 2011 Seven-year Wilkinson Microwave Anisotropy Probe (WMAP) observations: Cosmological interpretation *Astrophys. J. Suppl.* **192** 18 (*Preprint* arXiv:1001.4538)
Hinshaw G et al. [WMAP Collaboration] 2013 Nine-year Wilkinson Microwave Anisotropy Probe (WMAP) observations: Cosmological parameter results *Astrophys. J. Suppl.* **208** 19 (*Preprint* arXiv:1212.5226)
[3] Ade P A R et al. [Planck Collaboration] 2016 Planck 2015 results. XX. Constraints on inflation *Astron. Astrophys.* **594** A20 (*Preprint* arXiv:1502.02114)
Ade P A R et al. [Planck Collaboration] 2016 Planck 2015 results. XVI. Isotropy and statistics of the CMB *Astron. Astrophys.* **594** A16 (*Preprint* arXiv:1506.07135)
[4] Ellis G F R and MacCallum M A H 1969 A Class of homogeneous cosmological models *Commun. Math. Phys.* **12** 108
Ellis G F R 2006 The Bianchi models: Then and now *Gen. Rel. Grav.* **38** 1003
[5] Ade P A R et al. [Planck Collaboration] 2014 Planck 2013 results. XXVI. Background geometry and topology of the Universe *Astron. Astrophys.* **571** A26 (*Preprint* arXiv:1303.5086)

Ade P A R et al. [Planck Collaboration] 2016 Planck 2015 results. XVIII. Background geometry & topology *Astron. Astrophys.* **594** A18 (*Preprint* arXiv:1502.01593)

[6] Pitrou C, Pereira T S and Uzan J P 2008 Predictions from an anisotropic inflationary era *J. Cosmol. Astropart. Phys.* JCAP04(2008)004 (*Preprint* arXiv:0801.3596)

Gumrukcuoglu A E, Contaldi C R and Peloso M 2007 Inflationary perturbations in anisotropic backgrounds and their imprint on the cosmic microwave background *J. Cosmol. Astropart. Phys.* JCAP07(2007)005 (*Preprint* arXiv:0707.4179)

[7] Gibbons G W and Hawking S W 1977 Cosmological event horizons, thermodynamics, and particle creation *Phys. Rev.* D **15** 2738

Hawking S W and Moss I G 1982 Supercooled phase transitions in the very early universe *Phys. Lett.* B **110** 35

[8] Barrow J D 1987 Cosmic no hair theorems and inflation *Phys. Lett.* B **187** 12

[9] Wald R M 1983 Asymptotic behavior of homogeneous cosmological models in the presence of a positive cosmological constant *Phys. Rev.* D **28** 2118

Kleban M and Senatore L 2016 Inhomogeneous anisotropic cosmology *J. Cosmol. Astropart. Phys.* JCAP10(2016)022 (*Preprint* arXiv:1602.03520)

East W E, Kleban M, Linde A and Senatore L 2016 Beginning inflation in an inhomogeneous universe *J. Cosmol. Astropart. Phys.* JCAP09(2016)010 (*Preprint* arXiv:1511.05143)

Carroll S M and Chatwin-Davies A 2017 Cosmic equilibration: A holographic no-hair theorem from the generalized second law *Preprint* arXiv:1703.09241

[10] Kanno S, Soda J and Watanabe M a 2010 Anisotropic power-law inflation *J. Cosmol. Astropart. Phys.* JCAP12(2010)024 (*Preprint* arXiv:1010.5307)

Watanabe M a, Kanno S and Soda J 2009 Inflationary Universe with Anisotropic Hair *Phys. Rev. Lett.* **102** 191302 (*Preprint* arXiv:0902.2833)

Soda J 2015 Anisotropic Power-law Inflation: A counter example to the cosmic no-hair conjecture *J. Phys.: Conf. Series* **600** 012026 (*Preprint* arXiv:1410.8643)

[11] Maleknejad A, Sheikh-Jabbari M M and Soda J 2013 Gauge fields and inflation *Phys. Rept.* **528** 161 (*Preprint* arXiv:1212.2921)

Soda J 2012 Statistical anisotropy from anisotropic inflation *Classical Quantum Gravity* **29** 083001 (*Preprint* arXiv:1201.6434)

[12] Silverstein E and Tong D 2004 Scalar speed limits and cosmology: Acceleration from D-cceleration *Phys. Rev.* D **70** 103505 (*Preprint* hep-th/0310221)

Alishahiha M, Silverstein E and Tong D 2004 DBI in the sky: Non-Gaussianity from inflation with a speed limit *Phys. Rev.* D **70** 123505 (*Preprint* hep-th/0404084)

[13] Sasaki S, Yamaguchi M and Yokoyama D 2012 Supersymmetric DBI inflation *Phys. Lett.* B **718** 1 (*Preprint* arXiv:1205.1353)

[14] Deffayet C, Esposito-Farese G and Vikman A 2009 Covariant Galileon *Phys. Rev.* D **79** 084003 (*Preprint* arXiv:0901.1314)

Deffayet C, Deser S and Esposito-Farese G 2009 Generalized Galileons: All scalar models whose curved background extensions maintain second-order field equations and stress-tensors *Phys. Rev.* D **80** 064015 (*Preprint* arXiv:0906.1967)

Kobayashi T, Yamaguchi M and Yokoyama J 2010 Inflation Driven by the Galileon Field *Phys. Rev. Lett.* **105** 231302 (*Preprint* arXiv:1008.0603)

Gao X and Steer D A 2011 Inflation and primordial non-Gaussianities of "generalized Galileons" *J. Cosmol. Astropart. Phys.* JCAP12(2011)019 (*Preprint* arXiv:1107.2642)

[15] Caldwell R R 2002 A phantom menace? Cosmological consequences of a dark energy component with super-negative equation of state *Phys. Lett.* B **545** 23 (*Preprint* astro-ph/9908168)

Cai Y F, Saridakis E N, Setare M R and Xia J Q 2010 Quintom cosmology: Theoretical implications and observations *Phys. Rept.* **493** 1 (*Preprint* arXiv:0909.2776)

[16] Do T Q, Kao W F and Lin I C 2011 Anisotropic power-law inflation for a two scalar fields model *Phys. Rev.* D **83** 123002

[17] Do T Q and Kao W F 2011 Anisotropic power-law inflation for the Dirac-Born-Infeld theory *Phys. Rev.* D **84** 123009

[18] Do T Q and Kao W F 2016 Anisotropic power-law solutions for a supersymmetry Dirac-Born-Infeld theory *Classical Quantum Gravity* **33** 085009

[19] Do T Q and Nguyen S H Q 2017 Anisotropic power-law inflation in a two-scalar-field model with a mixed kinetic term *Int. J. Mod. Phys.* D **26** 1750072 (*Preprint* arXiv:1702.08308)

[20] Do T Q and Kao W F 2017 Bianchi type I anisotropic power-law solutions for the Galileon models *Phys. Rev.* D **96** 023529

Chameleon mechanism in f(R) modified gravitation model of polynomial-exponential form

Vo Van On and Nguyen Ngoc

Group of Computational Physics, Faculty of Natural Sciences, University of Thu Dau Mot, Binh Duong, Viet Nam

E-mail: onvv@tdmu.edu.vn, ngocm0815035@gstudent.ctu.edu.vn

Abstract. In this paper, we first present briefly to Chameleon mechanism in modified gravity of f(R), then we apply it to modified gravity of polynomial exponential form to find constraints from solar system experiments to parameters of α and β of the model. Results show that the Chameleon mechanism sets strict constraints on α and β parameters as follows: $0 < \beta < 7.67 \times 10^{-15}$ and $0 < \alpha < 9 \times 10^{-75}$.

Key words: Chameleon mechanism; polynomial exponential modified gravity; solar system constraints.

1. Introduction

In 1998, Astrophysicists uncovered a surprise that the universe was accelerating rather than slowing down as expected [1,2,3]. To explain this event there are three main approaches. In the first approach, people think that the Einstein equation with cosmological constant is sufficient to present the acceleration of universe. However, detailed calculations show that the cosmological constant from theory is about 120 orders of magnitude greater than its experimental value.This is the problem of cosmological constant [4], so this approach is not supported by the majority of scientists. In the second approach, the acceleration of universe causes by fifth force field called Quintessence field [5]. However, to satisfy constraints from observations and experiments, artificial conditions are required on this force field, such as: the field potential must be very flat so that the field rolls very slowly into the potential hole; the mass of quintessence particle is very small about$10^{-31}eV$. In the third approach, people change Einstein's theory to describe the acceleration of universe today but do not need to dark energy. Modified gravity of f(R) belongs this class of theory. In the class of theories, action of Einstein-Hilbert R in general relativity theory is replaced by a function of f(R). Here f(R) is analytic function which satisfies some constraint conditions.The first authors to take this path are Soviet physicist Starobinski in the 1980s of the last century[6], followed by American physicist Sean Caroll[7]. These early approaches gave physicists more emotion and experience.

We also use a theory model in the class to research the evolution of universe. In our model, f(R) takes of polynomial-exponential form[8,9]

$$S = \frac{1}{2\kappa^2}\int f(R)\sqrt{-g}d^4x + S_M(g_{\mu\upsilon}, \psi); \quad (1)$$

where

$$f(R) = R + a + \frac{\alpha}{R^m}\left(1 + bR^2 + cR^3\right)e^{-\beta R^n}; \tag{2}$$

with

$$\alpha > 0, \beta > 0; m = n = 1; a = -2\Lambda; b = c = 1. \tag{3}$$

This is a small class in f(R) theory class but it also is quite general, and is also of interest to many cosmologists today. This class called the class of polynomial- exponential modified gravity. The paper is constructed as follows: in section 2, we introduce the Chameleon mechanism in the class of f(R) modified gravity; in section 3, we present the Chameleon mechanism in the f(R) modified gravity of polynomial exponential form, then we calculate constraints on parameters of the model; section 4 is discussions and conclusion.

2. The Chameleon mechanism in f(R) modified gravity

We consider a static spherically symmetric compact object, it causes a perturbation on de Sitter background universe. The metric is as follows

$$ds^2 = -[1 - 2\varphi(r)]\mathrm{d}t^2 + [1 + 2\psi(r)]\mathrm{d}r^2 + r^2 d\Omega^2; \tag{4}$$

where φ and ψ are post- Newtonian potentials satisfy

$$|\varphi(r)|, |\psi(r)| << 1. \tag{5}$$

We only note small scales such that

$$H_0.r << 1; \tag{6}$$

we expand Ricci scalar about constant curvature of de Sitter universe

$$R(r) = R_0 + R_1. \tag{7}$$

The parameter of γ in Parameterized Post-Newtonian formalism(PPN)[10], is introduced as follows

$$\gamma = -\varphi(r)/\Psi(r); \tag{8}$$

three following assumptions are required

- $f(R)$ is analytic at R_0;
- $mr << 1$, i.e the mass of scalar field is very small and it has acting scale is great than the scale of solar system
- the pressure inside the object can be ignored,i.e, $T = T_0 + T_1 \simeq -\rho$;

people prove that if the condition is satisfied, we have with an any f(R) theory

$$\gamma = -\varphi(r)/\Psi(r) = 1/2; \tag{9}$$

while experiments in solar system give [11]

$$|\gamma - 1| < 2.3 \times 10^{-5}; \tag{10}$$

It means $\gamma = 1$. This puts an end to f(R) modified theories if there is no Chameleon mechanism as follows: It is thought that the second request is not always satisfied because the scalar field's mass may be large due to it depends on the curvature of the universe i.e on matter density of the surrounding environment. This mechanism was first proposed in 2004 by two American physicists, J. Khoury and A. Weltman, to overcome the above difficulty[12]. The oscillation of neutrino particles seems to support this mechanism because the mass of neutrino particles depends on the environment around them. Below we give a shortly review to this mechanism. We consider the following action

$$S = \frac{1}{2\kappa^2}\int f(R)\sqrt{-g}d^4x + S_M(g_{\mu\upsilon}, \psi); \tag{11}$$

where g is the determinant of $g_{\mu\nu}$, f(R) is an unknown function of the scalar curvature R and S_m is the matter action. We use a conformal transformation

$$\overline{g}_{\mu\nu} = pg_{\mu\nu}; \tag{12}$$

$$\phi = \frac{1}{2\beta_1}\ln p; \tag{13}$$

where

$$p \equiv \frac{df}{dR} = f'(R), \quad \beta_1 = \sqrt{\frac{1}{6}}. \tag{14}$$

This transforms the above action in Jordan frame to the Einstein frame[13]

$$S = \frac{1}{2}\int d^4x\sqrt{-g}[\overline{R} - g^{\mu\nu}\partial_\mu\phi\partial_\nu\phi - 2V(\phi)] + S_M(\overline{g}_{\mu\upsilon}e^{-2\beta_1\phi}, \psi); \tag{15}$$

the scalar field ϕ in the Einstein frame has a self-interacting potential

$$V(\phi) = \frac{1}{2}e^{-2\beta_1\phi}\left\{r\left[p(\phi)\right] - e^{-2\beta_1\phi}f\left(r\left[p(\phi)\right]\right)\right\}; \tag{16}$$

where r(p) is a solution of the equation

$$f'\left(r\left[p\right]\right) - p = 0. \tag{17}$$

The conformal transformation makes the coupling of the scalar field with the matter sector. The strength β_1 of this coupling is the same for all types of matter field and it equals $\frac{1}{\sqrt{6}}$. In such case, the scalar potential has an important role for consistency with local gravity experiments in solar system. The potential of the scalar field has to attribute an effective mass to the field which has a strong dependence on material density of environment. A theory that has such dependence called a chameleon theory. In a chameleon theory of f(R), the scalar field can be heavy enough in the environment of laboratories on the Earth so that the local gravity constraints suppressed, while it can be light enough in the low density cosmological environment to makes the cosmic acceleration. Variation of the action (15) with respect to $g_{\mu\nu}$ and ϕ gives the field equations

$$\overline{G}_{\mu\upsilon} = \partial_\mu\phi\partial_\upsilon\phi - \frac{1}{2}\overline{g}_{\mu\upsilon}\partial_\gamma\phi\partial^\gamma\phi - V(\phi)\overline{g}_{\mu\upsilon} + \overline{T}_{\mu\upsilon}; \tag{18}$$

$$\overline{\phi} - \frac{dV}{d\phi} = -\beta_1\overline{T}; \tag{19}$$

where

$$\overline{T}_{\mu\upsilon} = \frac{-2}{\sqrt{-g}}\frac{\delta S_m}{\delta\overline{g}^{\mu\upsilon}}; \tag{20}$$

and

$$\overline{T} = \overline{g}^{\mu\upsilon}\overline{T}_{\mu\upsilon}. \tag{21}$$

Covariant differentiation of (18) and Bianchi identities give

$$\overline{\nabla}^{\mu}\overline{T}_{\mu\upsilon} = \beta_1 \overline{T}\partial_{\upsilon}\phi; \tag{22}$$

Equation(20) indicates that the matter field is not generally conserved and feels a new force because of gradient of the scalar field. We now consider $T_{\mu\nu}$ as the stress-tensor of dust with energy density ρ in the Einstein frame. In a static spherically symmetric space-time region, equation (19) gives

$$\frac{d^2\phi}{d\bar{r}^2} + \frac{2}{\bar{r}}\frac{d\phi}{d\bar{r}} = \frac{dV_{eff}(\phi)}{d\phi}; \tag{23}$$

where $\bar{r}$ is the distance from the center of symmetric body to point that we investigated in Einstein frame. We have

$$V_{eff}(\phi) = V(\phi) - \frac{1}{4}\rho e^{-4\beta_1\phi}; \tag{24}$$

where $\overline{\rho} = e^{-4\beta_1\phi}\rho$ is the relation that relates between the energy density in Jordan and Einstein frames. We now consider a spherically symmetric body which has the radius $\bar{r_c}$ and assume that the energy density is constant $\overline{\rho}_{in}(\overline{r} < r_c)$ inside the body. We call the energy density outside the body by $\overline{\rho}_{out}(\overline{r} > r_c)$. We denote the values at two minimums of the effective potential $V_{eff}(\phi)$ inside and outside by φ_{in} and φ_{out}, respectively. They satisfy relations $V'_{eff}(\varphi_{in}) = 0$ and $V'_{eff}(\varphi_{out}) = 0$, where prime indicates differentiation of $V_{eff}(\phi)$ with respect to the argument of ϕ. Masses of small fluctuations about these minimums are given by $m_{in} = \left[V''_{eff}(\varphi_{in})\right]^{\frac{1}{2}}$ *and* $m_{out} = \left[V''_{eff}(\varphi_{out})\right]^{\frac{1}{2}}$.
Which depend on matter density of surrounding environment. The scalar field can has short-range effects in the solar system ($m > 10^{-3}eV$ for $\lambda < 0.2mm$ to avoid constraints by solar system experiments) due to the large material density in the solar system while it has a very small mass on cosmic scale due to very small density of universe and causes the current cosmic acceleration.
For dense compact objects which satisfy the thin shell following

$$\frac{\Delta\overline{r}_c}{\overline{r}_c} = \frac{\varphi_{out} - \varphi_{in}}{6\beta_1\phi_c} << 1; \tag{25}$$

where $\phi_c = \frac{M_c}{8\pi\overline{r}_c}$ is Newtonian potential at $\overline{r} = \overline{r}_c$, with M_c is the mass of the object. In this case, equation (23) with some appropriate boundary conditions gives the field profile outside the body[14]

$$\phi(\overline{r}) = -\frac{\beta_1}{4\pi}\frac{3\Delta\overline{r}_c}{\overline{r}_c}\frac{M_c e^{-m_{out}(\overline{r}-\overline{r}_c)}}{\overline{r}} + \varphi_{out}. \tag{26}$$

3. The Chameleon mechanism in f(R) modified gravity of polynomial-exponential form

We consider the above action

$$S = \frac{1}{2\kappa^2}\int f(R)\sqrt{-g}d^4x + S_M(g_{\mu\upsilon}, \psi); \tag{27}$$

with

$$f(R) = R - 2\Lambda + \frac{\alpha}{R}(1 + R^2 + R^3)e^{-\beta R}; \tag{28}$$

where g is determinant of $g_{\mu\nu}$, S_m is the action of matter which depends on the metric $g_{\mu\nu}$ and an any scalar field ψ.
Transforming this action from Jordan frame to Einstein frame as equations(12),(13),(14),(15),(16),(17) and assuming that $\phi << 1$, one can find the solution of $V'_{eff}(\phi) = 0$, We have

$$\begin{aligned} V'_{eff} &= V(\phi) - \tfrac{1}{4}\rho e^{-4\beta_1\phi}; \\ &= \frac{1}{2(1+4\beta_1\phi)}\left\{\frac{(8\alpha\beta\beta_1\phi-4\alpha\beta^2\beta_1\phi)-8\beta^2\beta_1^2\phi^2+8\Lambda\alpha\beta^2+6\alpha\beta\beta_1\phi+20\alpha^2\beta\beta_1\phi+\alpha^2\beta-2\alpha^2}{4\alpha\beta^2}\right\}; \end{aligned} \tag{29}$$

from

$$V'_{eff} = 0; \tag{30}$$

we have

$$\phi = \frac{-2\alpha\beta + 7\alpha}{8\beta\beta_1} \pm \sqrt{\frac{33\alpha^2 - 16\rho\alpha\beta^2}{64\beta^2\beta_1^2}}. \tag{31}$$

In section 3, we shall consider the thin-shell condition together constraints from the equivalence principle and fifth force experiments on the model

3.1. The thin shell condition
For a heavy and large object, the chameleon field is trapped inside the object so that its effect on other objects is only due to a thin shell near the outer surface of the object. The criterion for thin shell is given by equation of (25). When combining (25) and(31), we have

$$\begin{aligned} \frac{\Delta\bar{r}_c}{\bar{r}_c} &= \frac{\varphi_{out}-\varphi_{in}}{6\beta_1\Phi_c} << 1; \\ \frac{\Delta\bar{r}_c}{\bar{r}_c} &= \left(\pm\sqrt{\frac{33\alpha^2-16\rho_{out}\alpha\beta^2}{64\beta^2\beta_1^2}} \mp \sqrt{\frac{33\alpha^2-16\rho_{in}\alpha\beta^2}{64\beta^2\beta_1^2}}\right)\frac{1}{6\beta_1\Phi_c} << 1; \end{aligned} \tag{32}$$

where ρ_{in} and ρ_{out} are the energy densities inside and outside of the object in Jordan frame. In the weak approximation for a spherically symmetric object, the metric in Jordan frame has the form

$$ds^2 = -[1 - 2X(r)]\mathrm{d}t^2 + [1 + 2Y(r)]\mathrm{d}r^2 + r^2 d\Omega^2; \tag{33}$$

where $X(r)$ and $Y(r)$ are functions of the variable of r. The relation of r and $\overline{r}$ is $\overline{r} = p^{1/2}r$. We assume that $m_{out}.r << 1$ i.e, the Compton wave length m_{out}^{-1} is much large than solar system scales, then we have $\overline{r} \approx r$. In this case, Chameleon mechanism gives the parameter γ in Post-Newtonian formalism as follows

$$\gamma = \frac{3 - \frac{\Delta r_c}{r_c}}{3 + \frac{\Delta r_c}{r_c}} \simeq 1 - \frac{2}{3}\frac{\Delta r_c}{r_c}. \tag{34}$$

We obtain the thin shell condition for the Earth when applying (29). We consider the Earth as a solid sphere with the radius of $R_e = 6.4 \times 10^8$ cm and mean density $\rho_{out} \sim 10g/cm^3$. We also consider the Earth is surrounded by an atmosphere layer with homogeneous density $\rho_a \sim 10g/cm^3$ and thickness of 100km. When applying equation (32) and the constraint comes from Cassini tracking[15]

$$|1 - \gamma| < 2.3 \times 10^{-5}; \tag{35}$$

we have

$$\frac{\Delta R_e}{R_e} = \left(\sqrt{\frac{33\alpha^2 - 16\rho_{out}\alpha\beta^2}{64\beta^2\beta_1^2}} - \sqrt{\frac{33\alpha^2 - 16\rho_{in}\alpha\beta^2}{64\beta^2\beta_1^2}}\right)\frac{1}{6\beta_1\Phi_e} < 3.45 \times 10^{-5}; \tag{36}$$

or

$$\sqrt{\frac{33\alpha^2}{16\alpha\beta^2}-\rho_{out}}-\sqrt{\frac{33\alpha^2}{16\alpha\beta^2}-\rho_{in}}-\frac{165.6\beta_1^2\Phi_e\times 10^{-5}}{4\sqrt{\alpha}}<0; \tag{37}$$

set $a=\frac{33\alpha^2}{16\alpha\beta^2}$, $b=\frac{165.6\beta_1^2\Phi_e\times10^{-5}}{4\sqrt{\alpha}}$;

therefore

$$\sqrt{a-\rho_{out}}-\sqrt{a-\rho_{in}}-b<0; \tag{38}$$

when maintaining only first order of small parameter β, we have

$$\frac{256\beta^2}{33\times(165.6)^2\beta_1^4{\Phi_e}^2\times10^{-10}}<\frac{4}{(\rho_{in}-\rho_{out})^2}\Rightarrow\beta<\left(\frac{4.33(165.6)^2\beta_1^4{\Phi_e}^2\times10^{-10}}{256(\rho_{in}-\rho_{out})^2}\right)^{\frac{1}{2}}; \tag{39}$$

$$\beta<\left(\frac{4.33(165.6)^2\frac{1}{36}(6.25\times10^7)^2\times10^{-10}}{256(10^4)^2}\right)^{\frac{1}{2}}=1.23. \tag{40}$$

This is the constraint condition on the parameter β of the model. We see that this condition is not too strict on β due to in the model we have assumed from the beginning that $0<\beta<1$.

3.2. The equivalence principle

In this section, we consider constraints on the parameter β coming from possible violation of the equivalence principle. We assume that the Earth together its atmosphere is an isolated system. Far away from the Earth, the matter density is assumed to be homogeneous with the density $\rho_G\sim10^{-24}gr/cm^3$. We now consider conditions for the Earth to satisfy the thin shell condition. We take the thickness of atmosphere to be the thickness of thin shell about $\Delta R_a\sim100km$ and $R_a\sim4000km$, we obtain $\frac{\Delta R_a}{R_a}<1.5\times10^{-2}$. We also have relations

$$\frac{\Delta R_e}{R_e}=\frac{\varphi_a-\varphi_e}{6\beta\varphi_e}; \tag{41}$$

and

$$\frac{\Delta R_a}{R_a}=\frac{\varphi_G-\varphi_a}{6\beta\varphi_a}; \tag{42}$$

where φ_a , φ_e and φ_G are field values at local minimums of effective potential in regions $r<R_e$, $R_a>r>R_e$ and $r>R_a$ respectively. Because of the Newtonian potential inside a spherically symmetric object with the mass density ρ is $\Phi\ \propto\ \rho R^2$, we have $\Phi_e=10^4\Phi_a$. Where Φ_a and Φ_e are Newtonian potentials on the surface of the Earth and atmosphere, respectively. This gives

$$\frac{\Delta R_e}{R_e}\approx10^{-4}\frac{\Delta R_a}{R_a}; \tag{43}$$

such that, the condition for atmosphere to have a thin shell is

$$\frac{\Delta R_e}{R_e}<1.5\times10^{-6}. \tag{44}$$

The tests of the equivalence principle measure the difference of free fall accelerations of the Earth and the Moon towards to the Sun. The constraint on this difference of two accelerations is given by the relation[16]

$$\frac{|a_m-a_e|}{a_N}<10^{-13}; \tag{45}$$

IOP Conf. Series: Journal of Physics: Conf. Series **1034** (2018) 012002 doi:10.1088/1742-6596/1034/1/012002

where a_m and a_e are the accelerations of the Moon and the Earth, and a_N is Newtonian acceleration. The Sun and the Moon all have the thin shell and the field profile outside of these objects are given by formula (26) with respective quantities.The accelerations a_e and a_m are given by

$$a_e \approx a_N\{1 + 18\beta_1^2(\frac{\Delta R_e}{R_e})^2\frac{\Phi_e}{\Phi_s}\}; \tag{46}$$

$$a_m \approx a_N\{1 + 18\beta_1^2(\frac{\Delta R_e}{R_e})^2\frac{\Phi_e{}^2}{\Phi_s\Phi_m}\}; \tag{47}$$

where $\Phi_e = 6.9 \times 10^{-10}$, $\Phi_m = 3.14 \times 10^{-11}$ and $\Phi_s = 2.12 \times 10^{-6}$ are Newtonian potentials on the surface of the Earth, the Moon , and the Sun respectively. When differentiating two free - fall accelerations, we have

$$\frac{|a_m - a_e|}{a_N} = (0.13)\beta_1^2\left(\frac{\Delta R_e}{R_e}\right)^2; \tag{48}$$

combining with (45)and(48), we obtain result

$$\frac{\Delta R_e}{R_e} < 2.14 \times 10^{-6}; \tag{49}$$

therefore

$$\beta < 0.076. \tag{50}$$

This value is smaller than that in equation(40) about two orders of magnitude.

3.3. Fifth force

A fifth force is parameterized by a Yukawa potential following

$$U(r) = -\varepsilon\frac{m_1 m_2}{8\pi}\frac{e^{-r/\lambda}}{r}; \tag{51}$$

where m_1, m_2 are masses of two bodies, r is distance between them, ε is the strength of the interaction and λ is the range which is of the order of the size of vacuum chamber[17], namely $\lambda \approx R_{vac}$. The tightest bound on the strength of interaction is $\varepsilon < 10^{-3}$ [18]. Inside the chamber, one considers two identical bodies with uniform densities ρ_c, radius r_c and masses m_c. If two these bodies satisfy thin shell condition, their field profile outside these bodies are given by

$$\phi(r) = \frac{\beta_1}{4\pi}\frac{3\Delta r_e}{r_e}\frac{m_c e^{-r/R_{vac}}}{r} + \varphi_{vac}; \tag{52}$$

the interaction potential is

$$V(r) = -2\beta_1^2(\frac{3\Delta r_c}{r_c})^2\frac{m_c{}^2 e^{-r/R_{vac}}}{8\pi r}; \tag{53}$$

the bound on the strength of interaction becomes

$$2\beta_1^2(\frac{3\Delta r_e}{r_e})^2 < 10^{-3}; \tag{54}$$

combining(54) and (32), we have

$$\frac{\Delta R_e}{R_e} = \left(\sqrt{\frac{33\alpha^2 - 16\rho_{out}\alpha\beta^2}{64\beta^2\beta_1^2}} - \sqrt{\frac{33\alpha^2 - 16\rho_{in}\alpha\beta^2}{64\beta^2\beta_1^2}}\right)\frac{1}{6\beta_1\Phi_e} < \frac{10^{-3/2}}{3\sqrt{2}\beta_1}; \tag{55}$$

$$\beta < \left(\frac{4 \times 33\beta_1^2 {\Phi_e}^2 \times 10^{-3}}{2.(\rho_{in} - \rho_{out})^2} \right)^{\frac{1}{2}}; \tag{56}$$

$$\beta < \left(\frac{4 \times 33 \times \frac{1}{6}(6.95 \times 10^{-10})^2 \times 10^{-3}}{2 \times (9.5 \times 10^3)^2} \right)^{\frac{1}{2}} = 7.67 \times 10^{-15}. \tag{57}$$

.

3.4. The constraint on the parameter of alpha

From experiments in solar system, Gu and Lin[19] have obtained the constraints on Chameleon theories of f(R) as follows

$$-10^{-15} \leq F_R \leq 0; \tag{58}$$

with $R \sim 3 \times 10^5 H_0^2$ and

$$0 \leq RF_{RR} \leq 0.4; \tag{59}$$

for

$$R \geq 3 \times 10^5 H_0^2; \tag{60}$$

where $F(R) = f(R) - R = 2\Lambda + \frac{\alpha}{R}(1 + R^2 + R^3)e^{-\beta R}$ and $F_R = dF/dR; F_{RR} = d^2F/dR^2$; with

$$H_0^2 \approx 10^{-35} m^{-2} \Rightarrow R \approx 3 \times 10^{-30} m^{-2}. \tag{61}$$

.

In equation (28), we have approximatelly

$$e^{-\beta R} = 1 - \beta R; \tag{62}$$

therefore

$$F_R = \alpha \left[-3\beta R^2 - \frac{1}{R^2} - 2(\beta - 1) R + 1 \right]; \tag{63}$$

$$F_{RR} = 2\alpha \left[-\beta + \frac{1}{R^3} - 3\beta R + 1 \right]. \tag{64}$$

Since $R << 1$, therefore in expressions(63),(64), we have

$$F_R \cong -\frac{\alpha}{R^2}; \tag{65}$$

$$F_{RR} \cong +\frac{2\alpha}{R^3}; \tag{66}$$

when combining (65),(61) and (58), we have

$$\alpha \leq 9 \times 10^{-75}; \tag{67}$$

from (66),(61),(60) and (59), we have

$$\alpha \leq 1.8 \times 10^{-60}; \tag{68}$$

thus, the constraint on the parameter of alpha is

$$\alpha \leq 9 \times 10^{-75}. \tag{69}$$

From the constraints on beta parameter in formula (57) and on the alpha parameter in the formula (69), we see that these parameters are very small. Therefore we expect that the difference between this model and General relativity theory is observed only in regions that cosmic curvature is very large as in very early time of universe[20] and in regions of edge of black holes. In solar system, where the cosmic curvature is not large, predictive results from this model and Einstein's theory are indistinguishable.

4. Conclusion

In this paper, we have obtained constraints on the parameters of α and β of the model based on solar system experiments. The constraints on the parameter β required by the experiments of the thin shell condition of the Earth and the Equivalence principle are not too rigorous. The requirement from the absence of the fifth force caused by the scalar field puts a quite strict constraint on the parameter β and this is also the general status of many f(R) gravity modified models for this experiment. The constraint on the parameter α from experiments in solar system is rigorous. The above constraints are only based on experiments in the solar system. There are indications that the constraints from the inflation of the universe to these parameters are even more severe, which will be examined again in a near future article.

References

[1] Perlmutter S et al 1997 *Bull. Am. Astron. Soc.* **29** 1351
[2] Riess A G et al 1998 *Astron. J.* **116** 1009
[3] Perlmutter S et al 1999 *Astrophys. J.* **517** 565
[4] Weinberg S 1989 *Rev. Mod. Phys.* **61** 1 1-23
[5] Zlatev I, Wang L, Steinhardt P 1999 *Phys. Rev. Lett.* **82** 5 896-899
[6] Starobinsky A A 1980 *Phys. Lett.* B **91** 99-102
[7] Carroll S M, De Felice A, Duvvuri V, Easson D A, Trodden M, Turner M S 2005 *Phys. Rev.* D **71** 063513
[8] Vo Van On, Tran Trong Nguyen 2012 *Journal of Thu Dau Mot University* **6** 4 3-7
[9] Vo Van On, Tran Trong Nguyen 2015 *Chinese Journal of Physics* **53** 3 060101-1-060101-10
[10] Will C M 1971 *Astrophys. J.* **163** 611-628
[11] Faulkner T, Tegmark M, Bunn E F and Mao Y 2007 *Phys. Rev.* D **76** 063505
[12] Khoury J and Weltman A 2004 *Phys. Rev.*D **69** 044026
[13] Flanagan E E 2004 *Class. Q. Grav* **21** 3871
[14] Bisabr Y 2010 *Phys. Lett.* B **683** 96-100
[15] Will C M 2005 *Liv. Rev. Rel.* **9** 3
[16] Weinberg S 1972 *Gravitation and Cosmology* (John Wiley and Sons, New York)
[17] Khoury J and Weltman A 2004 *Phys. Rev. Lett.* **93** 171104
[18] Hoskins J K, Newman R D, Spero R and Schultz J 1985 *Phys. Rev.* D **32** 3084
[19] Lin W T, Gu J A, and Chen P 2010 Cosmological and Solar-System Tests of f(R) Modified Gravity (*Preprint astro-phy.CO*/1009.3488)
[20] Vo Van On, Truong Huu Nghi 2017 *Scientific Journal of Thu Dau Mot University* **34** 3 3-14

Dependence of Short-Channel Effects on Semiconductor Bandgap in Tunnel Field-Effect Transistors

Nguyen Dang Chien[1,2], Chun-Hsing Shih[3], Hung-Jin Teng[3,4], Cong-Kha Pham[5]

[1] Department for Management of Science and Technology Development, Ton Duc Thang University, Ho Chi Minh City, Viet Nam
[2] Faculty of Electrical & Electronics Engineering, Ton Duc Thang University, Ho Chi Minh City, Viet Nam
[3] Department of Electrical Engineering, National Chi Nan University, Nantou 54561, Taiwan
[4] Institute of Electronics Engineering, National Tsing Hua University, Hsinchu 30013, Taiwan
[5] Department of Network Engineering and Informatics, The University of Electro-Communications, Tokyo 182-8585, Japan

Email: nguyendangchien@tdt.edu.vn

Abstract. Scaling down the bandgap is considered as an essential approach to enhance the performance of tunnel field-effect transistors (TFETs). Using two-dimensional simulations, this study examines the dependence of short-channel effects on the semiconductor bandgap in TFETs. It is shown that the short-channel effect is more severe with using lower bandgap materials although the supply voltage is scaled in parallel with the bandgap. For a given bandgap material, the short-channel effect can be well evaluated by the increase of drain-induced barrier thinning (DIBT) with decreasing the channel length. For different bandgap TFETs, however, their short-channel effects cannot be compared properly by comparing the DIBTs. Adequately considering the effect of bandgap on the TFET scalability is necessary in designing scaled integrated circuits.

1. Introduction

Tunnel field-effect transistors (TFETs) with a steep subthreshold swing of sub-60 mV/decade [1], [2] have attracted much attention since the demand of low power consumption becomes a key factor in designing electronic devices. However, the conduction mechanism based on band-to-band tunneling (BTBT) does not allow the TFET on-current as high as that of conventional MOSFETs [3]. Over the past decade, many advanced techniques involving new materials and structures have been suggested to ameliorate the on-current of TFETs [4]-[7]. Among proposed solutions, using low-bandgap materials has been demonstrated as the most effective method to improve the on-current [4] since the tunneling probability depends exponentially on the bandgap which is treated as the height of tunnel barrier [8].

With the continuous trend of decreasing the feature size of transistors to keep the pace of Moore's law, the length scaling of TFETs is also one of the most important problems that need to be addressed. Since TFETs are operated in reverse-biased condition, valence electrons in the source can directly tunnel to the drain at off-state to cause severe short-channel effects in sub-30 nm uniform TFETs [9], [10]. Some sophisticated techniques have been developed to scale TFETs into sub-10 nm regimes

IOP Conf. Series: Journal of Physics: Conf. Series **1034** (2018) 012003 doi:10.1088/1742-6596/1034/1/012003

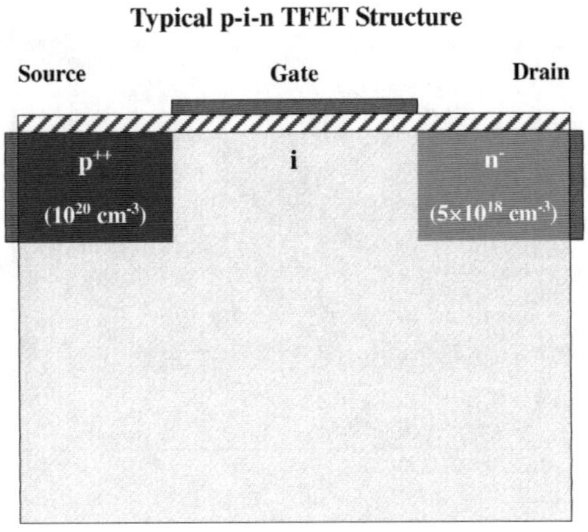

Figure 1. Typical p-i-n structure of tunnel field-effect transistors used in this study.

[11]-[14]. While the scaling of bandgap is critical to boost the on-current, its effects on the short-channel effect have not been clearly understood. Our recent work has pointed out that the bandgap parameter should be taken into account in studying the short-channel effect of TFETs [12]. Therefore, an adequate understanding on the role of bandgap in determining the short-channel performance is important in applying low-bandgap materials for scaled TFETs.

In this paper, we examined the dependence of short-channel effects on the bandgap of semiconductors in TFET devices. Two-dimensional simulations [15] based on the nonlocal approach of direct and indirect Kane's BTBT models [8] were performed to produce their electrical characteristics. The total tunneling current in TFETs was determined by the BTBT generation rate (G_{BTBT}) which was modeled in the Kane's formalism as:

$$G_{\mathrm{BTBT}} = A\frac{\xi^{\gamma}}{E_g^{1/2}}\exp(-B\frac{E_g^{3/2}}{\xi}), \tag{1}$$

where E_g is the semiconductor bandgap, is the nonlocal electric field which is the average of electric field along the tunnel path. In the formula of tunneling rate (1), $\gamma = 2$ was applied for direct tunneling and $\gamma = 2.5$ was used for indirect tunneling. Input factors A and B depend on material parameters and their values can be manually adjusted in the simulator. To investigate the bandgap dependence of short-channel effects, three different bandgap materials including high-bandgap Si (1.12 eV), low-bandgap Ge (0.8 eV) and InAs (0.36 eV) were adopted. These low-bandgap semiconductors were demonstrated as excellent candidates for TFET devices because of their CMOS-compatible fabrication technology [16]. The indirect BTBT model was used for Si-based TFETs, whereas the direct BTBT model was applied for Ge and InAs devices. In the simulations, we used the values of A and B that were theoretically calculated and experimentally verified for Si in [17], Ge and InAs in [18]. For more realistic simulations, the Fermi-Dirac distribution and the Shockley-Read-Hall recombination were also enabled. To exactly evaluate the bandgap dependence of short-channel effects, however, the process-sensitive trap-assisted tunneling and the bandgap narrowing due to heavy doping were not included.

2. Short-channel effects

Figure 1 shows the schematic view of a typical p-i-n TFET used in this study. The uniform structure was utilized to exclude any effects of structure parameters on short-channel effects, which may create difficulties for the studying purpose. The source region was heavily doped by a p-type impurity of 10^{20} cm^{-3} to enhance the on-current, whereas a medium donor concentration of 5×10^{18} cm^{-3} was employed for the drain to diminish the ambipolar off-current [19]. A very low n-type doping concentration of

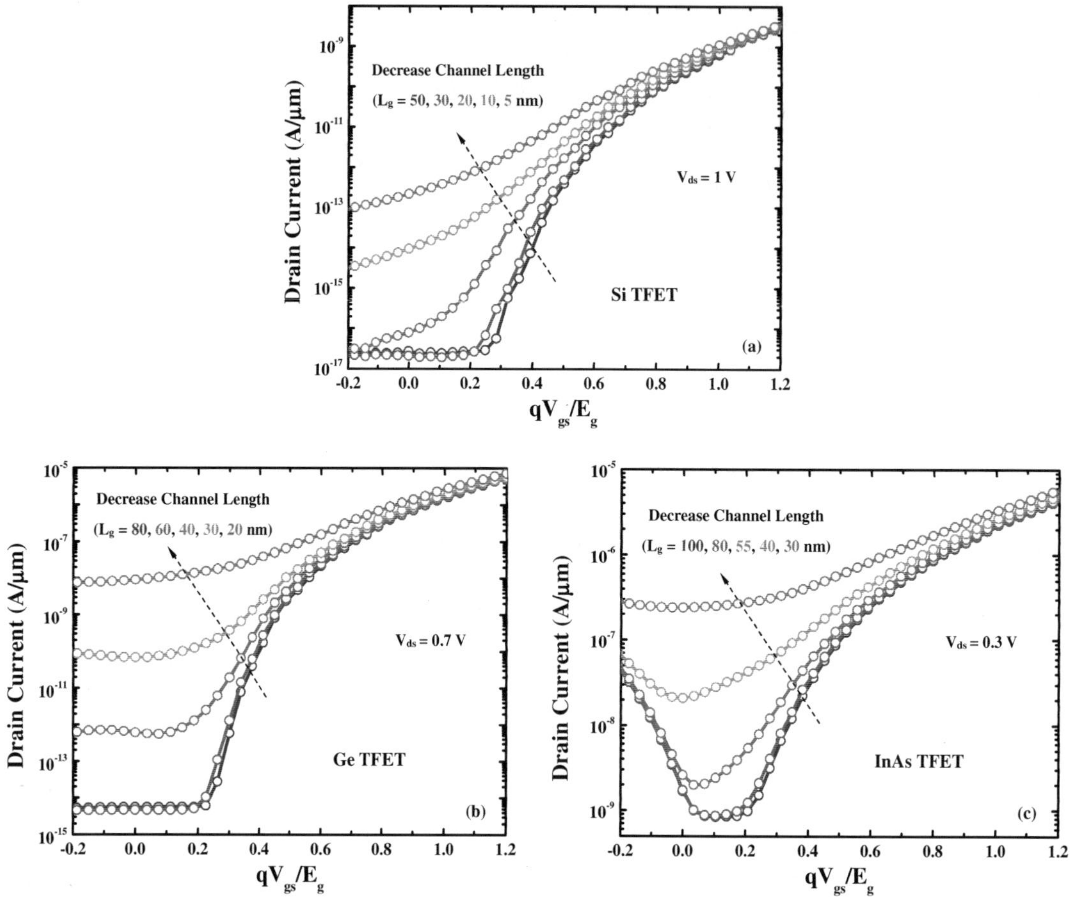

Figure 2. Current-voltage characteristics of TFET devices based on (a) Si, (b) Ge, and (c) InAs.

10^{15} cm^{-3} was defined for the channel and a practical doping gradient of 2 nm/decade was applied at source/drain junctions. A reasonable equivalent oxide thickness of 1 nm was specified for the gate insulator layer and the gate workfunction was set at 4.25 eV for all TFETs.

To investigate the dependence of short-channel effects on the bandgap, figure 2 shows the current-voltage characteristics of TFETs using Si, Ge and InAs. Because the drain voltage must be scaled in parallel with the bandgap to eliminate the ambipolar off-leakage [18], the supply voltage (V_{dd}) for drain in each case of material is slightly smaller than their bandgap voltage (E_g/q), namely 1, 0.7 and 0.3 V for Si, Ge and InAs TFETs, respectively. Moreover, setting the drain voltage in the scale of bandgap is also to ensure the appropriate and significant evaluation of short-channel effects because the smaller bandgap is suitable for lower power applications. For easily comparing short-channel effects in the three cases, the gate voltage is normalized by the bandgap voltage. Generally, the short-channel effects are clearly observed in all Si, Ge and InAs TFETs at very short channel lengths. In particular, the on-off characteristics start deteriorating because of substantial short-channel effects when the channel lengths are scaled below 30 nm for Si, 60 nm for Ge, and 80 nm for InAs TFETs. It means that with the same channel length of 40 nm, the short-channel effect is negligible in the Si TFET, but it is considerable in the Ge TFET and severe in the InAs TFET. Notably, the channel length limit, which separates the two regimes of negligible and severe short-channel effects, increases when going from high- to low-bandgap TFETs. To exactly explain the differences of the length limits, it is noted that there are only two factors that are different among the TFETs, including the bandgap and

IOP Conf. Series: Journal of Physics: Conf. Series **1034** (2018) 012003 doi:10.1088/1742-6596/1034/1/012003

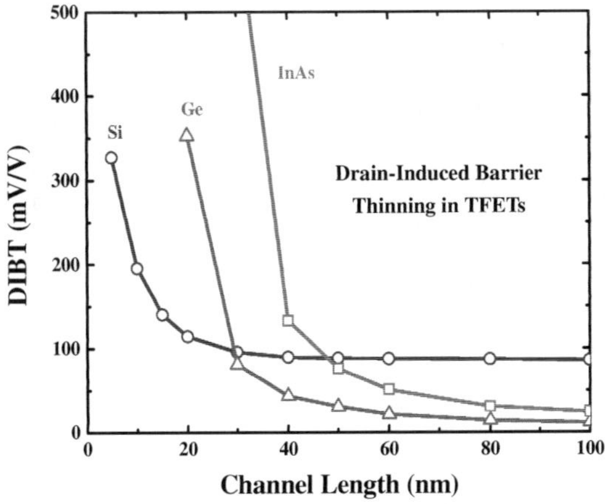

Figure 3. Drain-induced barrier thinning (DIBT) as a function of channel length for Si, Ge and InAs TFETs.

drain voltage. Logically, because decreasing the drain voltage helps to decrease the short-channel effect of TFETs [20], the increase in the length limit is obviously attributed to the decrease in the bandgap of semiconductors. Physically, the increase in short-channel effect with decreasing the bandgap is due to the exponential dependence of tunneling probability on the bandgap. More specifically, the increase of tunneling probability is much more rapid than the decrease of drain voltage when scaling down the bandgap.

3. Drain-induced barrier thinning

In traditional MOSFETs, if the channel length is short enough, the electric field of drain can penetrate deeply into the region of the source-channel junction. The partial merge of the source and drain fields leads to the lowering of thermal barrier to deteriorate the on-off characteristics of short-channel devices. This drain-induced barrier lowering (DIBL) can be used to directly measure the short-channel effect of MOSFETs because the larger the DIBL, the stronger the short-channel effect is. In TFET devices, similarly, the drain field can encroach upon the source-channel region to thin the tunnel barrier. Therefore, this drain-induced barrier thinning (DIBT) also degrades the on-off switching of TFETs. However, the current of TFETs is controlled not only by the tunnel barrier width, but also by the tunnel barrier height. Merely using DIBT is not enough to estimate the short-channel performance of TFETs [12]. In this section, the bandgap dependence of DIBT and its role in explaining the short-channel effect of TFETs are considered by investigating different bandgap materials.

Figure 3 shows the DIBT as a function of channel length for Si, Ge and InAs TFETs. For fair comparisons, the threshold voltage used in the DIBT calculation is defined separately for each case of material using the constant current method. The current level used for each material is determined by the 100nm TFET at the gate voltage of $0.4E_g/q$ higher than the onset voltage. The calculation of DIBT is based on the threshold voltages attracted at V_{dd} and $0.1V_{dd}$. Overall, the DIBT tends to increase as the channel length decreases for all materials. The DIBT starts increasing significantly when the channel lengths are decreased below 30 nm for Si, 60 nm for Ge, and 80 nm for InAs TFETs. These values of channel lengths are identical to those observed in figure 2. Importantly, only the variation trend of the DIBT, but not its magnitude, depends on the bandgap similarly to that of the short-channel effect. It implies that for a given semiconductor bandgap, the DIBT variation with scaling the channel length can be used as a good indicator to measure the short-channel effect of TFETs. However, the DIBT magnitude cannot be used to compare short-channel effects between different bandgap devices. As seen in the figure, for the same channel length of 40 nm, the DIBT of the Si TFET is considerably higher than that of the Ge TFET, whereas the short-channel effect of the Si TFET is much weaker than

IOP Conf. Series: Journal of Physics: Conf. Series **1034** (2018) 012003 doi:10.1088/1742-6596/1034/1/012003

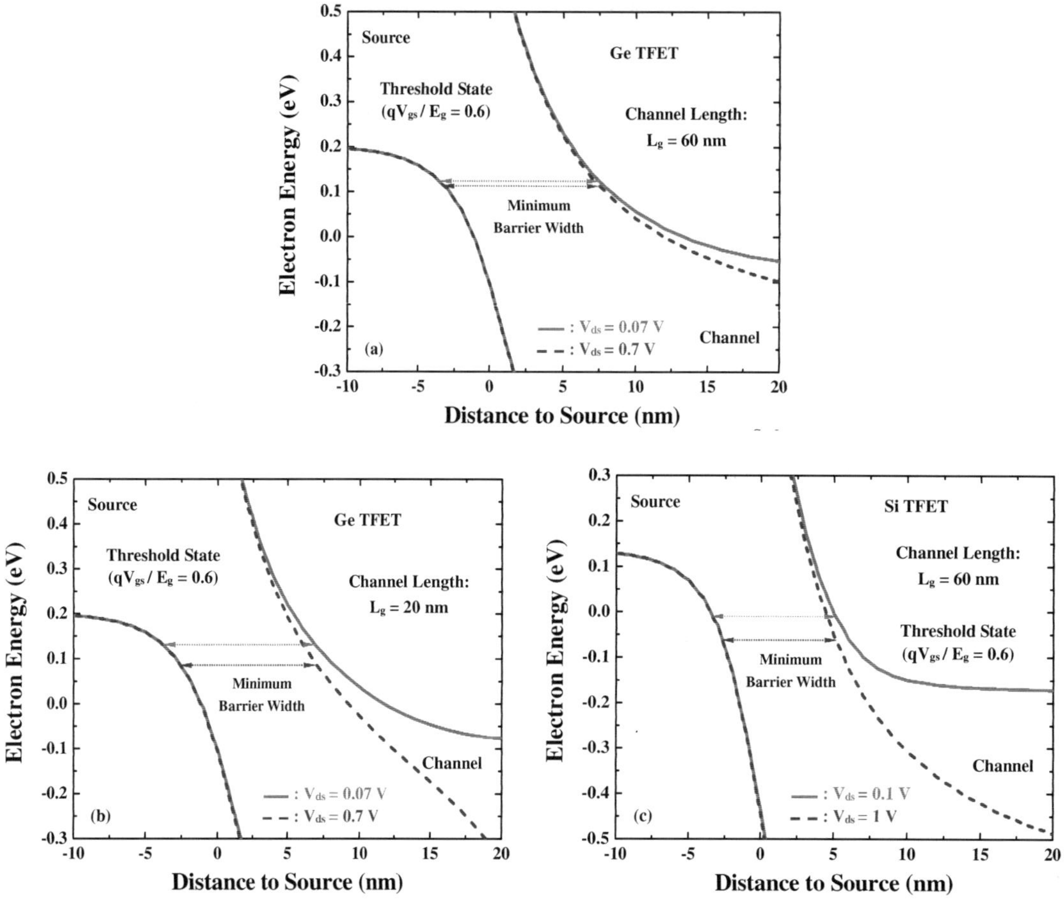

Figure 4. Energy-band diagrams at threshold state of (a) long-channel Ge, (b) short-channel Ge, and (c) long-channel Si TFETs.

that of the Ge TFET. For longer channel lengths, the DIBT of the Si TFET is even considerably greater than that of the Ge and InAs devices. Finally, it is seen at the short-channel regimes that the increase of DIBT is more rapid in TFETs with lower bandgap materials, which is a direct consequence of enhanced tunneling probability due to decreased bandgap. In other words, the short-channel performance of low-bandgap TFETs degrades at a faster rate than that of high-bandgap counterparts with aggressive channel length scaling.

In order to understand the increase of DIBT when scaling the channel length, figures 4(a)-(b) plot the energy-band diagrams at threshold state of Ge TFETs with different channel lengths. For the 60nm TFET, the conduction band at the region of the shortest tunnel path in case of $V_{ds} = 0.7$ V is very close to that in case of $V_{ds} = 0.07$ V because the drain field does not penetrate significantly into the channel. Therefore, the minimum barrier widths in the two cases are almost equal to produce a small DIBT. For the 20nm TFET, however, the deep encroachment of the drain field on the channel at the high drain voltage pulls down the channel conduction band significantly. Consequently, the shortest tunnel path regions are clearly separated and the minimum barrier width is appreciably decreased at the high drain voltage to get a large DIBT. Here, only the energy-band diagrams of Ge-based TFETs are exampled for illustrating the explanations because the similar mechanism of DIBT trends also occurs in higher-bandgap Si and lower-bandgap InAs TFETs. For explaining the large DIBT at long-channel regime in high-bandgap Si TFETs compared to low-bandgap Ge TFETs (and similarly to InAs TFETs), figure

4(c) shows the energy-band diagrams at threshold state of the 60nm Si TFET at low and high drain voltages. Although the long channel can effectively prevent the drain field encroachment on the channel, the high drain voltage associated with the small carrier density in the channel pulls the channel conduction band down largely. As a result, the minimum barrier width at the high voltage is much smaller than that at the low voltage, which results in the large DIBT.

4. Conclusion

The dependence of short-channel effects on the semiconductor bandgap in TFETs have been examined to show that the device scalability is poorer with using lower bandgap materials. For a given bandgap, the variation of drain-induced barrier thinning with scaling the channel length can serve as a good indicator to evaluate the short-channel effect of TFETs, but its magnitude cannot be used to compare short-channel effects between different bandgap devices. With the necessity of decreasing the bandgap for enhancing the on-current, the bandgap dependence of short-channel effects should be properly considered in scaling TFET devices.

Acknowledgments

This study is funded by Ton Duc Thang University, Vietnam. This work is also partially supported by the Ministry of Science and Technology and the National Center for High-Performance Computing of Taiwan.

References

[1] Appenzeller J, Lin Y-M, Knoch J and Avouris Ph 2004 *Phys. Rev. Lett.* **93** 196905.
[2] Choi W Y, Park B-G, Lee J D and Liu T-J K 2007 *IEEE Electron Device Lett.* **28** 743
[3] Seabaugh A C and Zhang Q 2010 *Proc. IEEE* **98** 2095
[4] Zhang Q, Fang T, Xing H, Seabaugh A and Jena D 2008 *IEEE Electron Device Lett.* **29** 1344
[5] Chien N D, Shih C-H and Vinh L T 2013 *J. Appl. Phys.* **114** 094507 [2013 *Erratum* **114** 189901]
[6] Ilatikhameneh H, Klimeck G, Appenzeller J and Rahman R 2016 *IEEE J. Electron Devices Soc.* **4** 260
[7] Wang P-Y and Tsui B-Y 2016 *IEEE Trans. Nanotechnol.* **15** 74
[8] Kane E O 1961 *J. Appl. Phys.* **31** 83
[9] Bardon M G, Neves H P, Puers R and Hoof C V 2010 *IEEE Trans. Electron Devices* **57** 827
[10] Liu L, Mohata D and Datta S 2012 *IEEE Trans. Electron Devices* **59** 902
[11] Yoon Y J, Eun H R, Seo J H, Kang H-S, Lee S M, Lee J, Cho S, Tae H-S, Lee J-H and Kang I M 2015 *J. Nanosci. Nanotechnol.* **15** 7430
[12] Chien N D and Shih C-H 2016 *Superlattice. Microst.* **100** 857
[13] Ilatikhameneh H, Klimeck G and Rahman R 2016 *IEEE Electron Device Lett.* **37** 115
[14] Wan J, Zaslavsky A, Le Royer C and Cristoloveanu S 2013 *IEEE Electron Device Lett.* **34** 24
[15] 2010 *Synopsys MEDICI User's Manual* (California: Synopsys Inc.)
[16] Kim S H, Kam H, Hu C, Liu T-J K 2009 *Proc. VLSI Symp. Tech. Dig.* (16-18 Jun 2009, Honolulu) p 178
[17] Shih C-H and Chien N D 2013 *J. Appl. Phys.* **113** 134507
[18] Shih C-H and Chien N D 2014 *IEEE Trans. Electron Devices* **61** 1907
[19] Toh E-H, Wang G H, Samudra G and Yeo Y-C 2008 *J. Appl. Phys.* **103** 104504
[20] Chien N D, Shih C-H, Chen Y-H and Thu N T 2016 *Proc. Int. Conf. on Electronics, Information and Communication* (27-30 Jan 2016, Da Nang) p 10

Improvement quantum teleportation via the pair coherent states

Tran Quang Dat[1,2], **Truong Minh Duc**[1], **Ho Sy Chuong**[1,3]

[1] Center for Theoretical and Computational Physics, College of Education, Hue University, Hue city, Vietnam

[2] University of Transport and Communications, Campus in Ho Chi Minh city, 450 Le Van Viet, 9 district, Ho Chi Minh city, Vietnam

[3] Dong Nai University, 4 Le Quy Don, Tan Hiep, Bien Hoa City, Vietnam

E-mail: tmduc2009@gmail.com

Abstract. In this paper, we present two methods to improve the average fidelity in the quantum teleportation processes to teleport a coherent state via the pair coherent states. We use the measuring of the orthogonal quadrature components in the first protocol and the photon number sum and phase difference in the second. The results show that the average fidelity in the first protocol is enhanced by performing a photon number shifting at a receiver, and F_{av} can approach to the unit if q becomes very big. In the second protocol, the average fidelity increases with increasing the amplitude of the pair coherent states. It approaches to the unit in the limit of the big amplitude values of the pair coherent states and small of the input state. The average fidelity of both methods is greatly improved comparing with the Gabris-Agarwal protocol and it approaches to unit depending on the parameters involved.

1. Introduction

The quantum teleportation is a disembodied information transfer process. After ideal Bennett's protocol [1], the quantum teleportation has developed further to many aspects. For example, the protocols were proposed by using the measuring of the orthogonal quadrature components [2], [3], the Bell measurement [4], the photon number difference and phase sum [5], the photon number sum and phase difference [6], and the multi-particle entanglement sources [7], [8].

According to the quantum teleportation processes, the nonclassical states play a very important role as the entanglement sources to teleport the arbitrary states [9], [10], [11], [12], [13], [14], [15]. One of the nonclassical states that we are interested in studying is the pair coherent state [16]. Recently, some states by generalization of the pair coherent states have been introduced and their nonclassical states have been studied [17], [18], [19]. The pair coherent states were defined by Agarwal as the eigenstates of the pair boson annihilation operators $\hat{a}\hat{b}$ and photon number operators difference $\hat{N}_b - \hat{N}_a = \hat{a}^+\hat{a} - \hat{b}^+\hat{b}$ in the form:

$$\hat{a}\hat{b}|\Psi_q\rangle_{ab} = \xi|\Psi_q\rangle_{ab},$$
$$(\hat{N}_b - \hat{N}_a)|\Psi_q\rangle_{ab} = q|\Psi_q\rangle_{ab}, \tag{1}$$

where ξ is complex, $\xi = |\xi|e^{i\varphi}$, $0 \leq \varphi \leq 2\pi$, q is positive integer. In Fock states, such states read

$$|\Psi_q\rangle_{ab} = \mathcal{N}_q(|\xi|) \sum_{n=0}^{\infty} \frac{\xi^n}{\sqrt{n!(n+q)!}} |n, n+q\rangle_{ab}, \tag{2}$$

where

$$\mathcal{N}_q(|\xi|) = \frac{1}{\sqrt{\xi^q I_q(2|\xi|)}}, \tag{3}$$

is the normalized factor, and $I_q(2|\xi|)$ is the modified Bessel function.

The quantum teleportation process by using Gabris-Agarwal protocol [11], in which the pair coherent states as the entanglement sources are feasible but the maximum of the average fidelity in the quantum teleportation process is only 0.75884. In this paper, we improve the quantum teleportation via the pair coherent states by using two different ways as the orthogonal quadrature components [3] and the photon number sum and phase difference protocols [6]. We show that in both ways, the average fidelity of the quantum teleportation process can be greatly improved and it approaches to the unit corresponding on the maximum fidelity depending on the appropriate parameters.

2. The quantum teleportation uses the orthogonal quadrature components protocol

The orthogonal quadrature components protocol was introduced by Furusawa *et al* [2] and developed more further by several authors [3], [5], [11], [14]. Using this protocol, we assume that A and B parties are shared by the pair coherent states that are given in Eqs. (2) and (3). The A party has mode a, and mode b is given for the B party. A wants to teleport to B a coherent state $|\alpha\rangle_c$ of mode c. To begin the process, A performs a combination of the three-mode state is

$$|\Phi_{in}\rangle_{abc} = |\Psi_q\rangle_{ab}|\alpha\rangle_c. \tag{4}$$

Next, he measures the orthogonal quadrature components [3] on modes a and c. After measurement, a non-normalized state of B becomes

$$|\Phi\rangle_b = \frac{e^{(\beta^*\alpha-\beta\alpha^*)/2}e^{-|\alpha-\beta|^2/2}}{\sqrt{\pi}} \sum_{n=0}^{\infty} \frac{c_{n,q}(\alpha-\beta)^n}{\sqrt{n!}} |n+q\rangle_b, \tag{5}$$

where

$$c_{n,q} = \mathcal{N}_q(|\xi|) \frac{\xi^n}{\sqrt{n!(n+q)!}}. \tag{6}$$

Then, A sends the measurement results to B by a classical channel. When B receives these data, first, he performs a photon number shifting by using operator [20]

$$\hat{U}_q = \sum_{j=0}^{\infty} |j\rangle\langle j+q|, \tag{7}$$

then, uses the displacement operator $\hat{D}(\beta)$, and the normalized output state reads

$$|\Phi_{nor.}\rangle_{out} = \frac{e^{(\beta^*\alpha-\beta\alpha^*)/2}e^{-|\alpha-\beta|^2/2}}{\sqrt{P(\beta)}\sqrt{\pi}} \sum_{n=0}^{\infty} \frac{c_{n,q}(\alpha-\beta)^n}{\sqrt{n!}} \hat{D}(\beta)|n\rangle, \tag{8}$$

where the probability of the measurement $P(\beta) = {}_b\langle\Phi|\Phi\rangle_b$ is in the form .

$$P(\beta) = \frac{e^{-|\alpha-\beta|^2}}{\pi}\sum_{n=0}^{\infty}\frac{|c_{n,q}|^2|\alpha-\beta|^{2n}}{n!}. \tag{9}$$

In case $\xi = |\xi|$, the average fidelity of the quantum teleportation process is

$$\begin{aligned} F_{av} &= \int |\langle\alpha|\Phi\rangle_{out}|^2 d^2\beta \\ &= \frac{\mathcal{N}_q^2(|\xi|)}{2}\sum_{m,n=0}^{\infty}\frac{|\xi/2|^{m+n}(m+n)!}{m!n!\sqrt{m!n!(m+q)!(n+q)!}}. \end{aligned} \tag{10}$$

3. The quantum teleportation uses the measuring of photon number sum and phase difference protocol

In the measuring of the photon number sum and phase difference protocol [6], the input state $|\alpha\rangle_c$ is expanded in Fock space as

$$|\alpha\rangle_c = \sum_{m=0}^{\infty} d_m|m\rangle_c, \tag{11}$$

where $|m\rangle_c$ is the Fock state, and

$$d_m = \frac{e^{-|\alpha|^2/2}\alpha^m}{\sqrt{m!}}. \tag{12}$$

We rewrite Eq. (4) as

$$|\Phi_{in}\rangle_{abc} = \sum_{m,n=0}^{\infty} c_{n,q}d_m|n, n+q, m\rangle_{abc}. \tag{13}$$

A measures the photon number sum and phase difference on two modes a and c, a state of the system becomes

$$|\Phi_b\rangle = \frac{{}_{ac}\langle\phi_N^-|\Phi_{in}\rangle_{abc}}{\sqrt{P}}, \tag{14}$$

where $P = \langle\Psi_b|\Psi_b\rangle$ is probability to obtain the photon numbers sum N and phase difference ϕ^-, and $|\phi_N^-\rangle_{ac}$ is the eigenstate of the photon number sum operator $\hat{N} = \hat{N}_a + \hat{N}_c$ and the phase difference operator $\hat{\phi}^- = \hat{\phi}_a - \hat{\phi}_c$. Such state reads [21]

$$|\phi_N^-\rangle_{ac} = \frac{1}{\sqrt{2\pi}}\sum_{j=0}^{N} e^{ij\phi^-}|j\rangle_a|N-j\rangle_c, \tag{15}$$

with ϕ^- is restricted in the windows $\phi_0^- \leq \phi^- < \phi_0^- + 2\pi$ with ϕ_0^- is an arbitrary real number. We clearly write the Eq. (14) as

$$|\Phi_b\rangle = \frac{P^{-1/2}}{\sqrt{2\pi}}\sum_{n=0}^{N} c_{n,q}d_{N-n}e^{-in\phi^-}|n+q\rangle_b, \tag{16}$$

with

$$P = \frac{1}{2\pi} \sum_{n=0}^{N} |c_{n,q}|^2 |d_{N-n}|^2. \tag{17}$$

After measurement, A sends to B the number of N and ϕ^- by a classical channel. Using these data, B rotates his phase by the unitary operator $\hat{U} = e^{i(\hat{N}_b - q)\phi^-}$ with $\hat{N}_b$ is the photon number operator of mode b. The state of B becomes

$$|\Phi\rangle = \frac{P^{-1/2}}{\sqrt{2\pi}} \sum_{n=0}^{N} c_{n,q} d_{N-n} |n+q\rangle_b. \tag{18}$$

Then, B transforms the photon number $n+q$ to become $N-n$. The teleportation process is completed. The output state is

$$|\Phi\rangle_{out} = \frac{P^{-1/2} e^{-|\alpha|^2/2}}{\sqrt{2\pi}} \sum_{n=0}^{N} c_{n,q} \frac{\alpha^{N-n}}{\sqrt{(N-n)!}} |N-n\rangle, \tag{19}$$

and in case $q = 0$ and $\xi = |\xi|$, the average fidelity is

$$F_{av} = \mathcal{N}_0^2(|\xi|) e^{-2|\alpha|^2} \sum_{N=0}^{\infty} \left| \sum_{n=0}^{N} \frac{|\xi|^n |\alpha|^{2N-2n}}{(N-n)!n!} \right|^2. \tag{20}$$

4. Results and discussions

Using the Eqs. (10) and (20), we discus about enhancing the average fidelity of the quantum teleportation process via pair coherent states as the entanglement sources in two difference protocols. Based on Eq. (10), it is easy to see that the average fidelity in the orthogonal quadrature components protocol does not depend on the amplitude of the input state $|\alpha|$, but depends on the amplitude of the pair coherent states $|\xi|$ and q. The figure 1 plots the dependence of the F_{av} as a function of $|\xi|$ and q corresponding to $q = 0, 1, 3$ and 6, respectively. The difference between our protocol and Gabris-Agarwal protocol [11] is that while the receiver in our protocol performs a photon number shifting by using operator in Eq. (7) and the receiver in Gabris-Agarwal protocol is not. That makes the average fidelity in the quantum teleportation process depends on q. Note that the case $q = 0$, the F_{av} in Eq. (10) reduces to the Gabris-Agarwal protocol, in which the highest average fidelity is 0.75884. It shows in figure 1 that the teleportation process is successful, and the average fidelity is more improved by increasing of q. If we add a photon number shifting in the quantum teleportation process, maximum of the average fidelity can approach to the unit by choosing high value of charge q. When a receiver performs a photon number shifting before using a displacement operator in the process to construct the output state, the average fidelity is enhanced by increasing of q, even F_{av} can approach to the unit if q becomes very big.

Based on Eq. (20), we can see that the average fidelity in the measuring of photon number sum and phase difference protocol depends on both the amplitudes of the input state $|\alpha|$ and the pair coherent states $|\xi|$. The figure 2 plots the dependence of the F_{av} as a function of $|\xi|$ and $|\alpha|$ corresponding to $|\alpha| = 0.5$, 1 and 2, respectively. It shows that for giving the values of $|\alpha|$, the quantum teleportation process is also successful, and the average fidelity is more increased by increasing of the $|\xi|$. As previous discussion, the maximum of the average fidelity in the quantum teleportation process in the Gabris-Agarwal protocol [11] is only 0.75884. But in the measuring of photon number sum and phase difference protocol, the average fidelity can approach to the

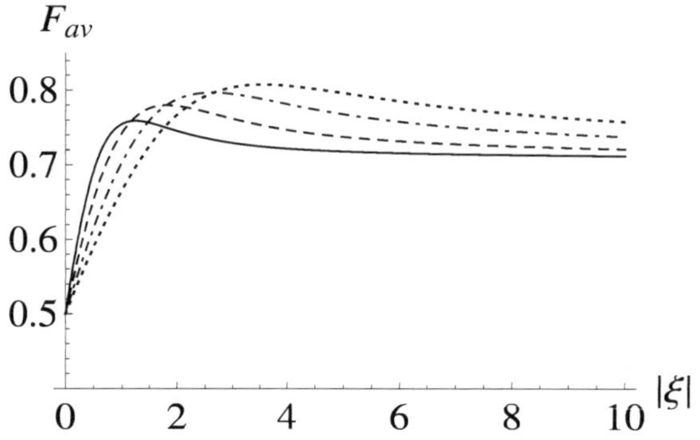

Figure 1. F_{av} as a function of $|\xi|$ and q, from bottom (solid lines) to top (dotted lines) corresponding to $q = 0, 1, 3$ and 6, respectively.

unit by choosing small value of the input state $|\alpha|$ and the big value of the amplitude of the pair coherent states $|\xi|$. For example, if $|\alpha| = 0.5$ and $|\xi| = 15$, the average fidelity of photon number sum and phase difference protocol is 0.99169. That means the quantum teleportation process can be greatly improved and it approaches to maximum value. For given the amplitude of the input state $|\alpha|$, the improvement of F_{av} by increasing of the amplitude of the pair coherent states is achieved, i.e., the average fidelity increases with the increasing of $|\xi|$. With a small value of the input state $|\alpha|$, the average fidelity can approach to the unit in the limit of the big amplitude value of the pair coherent states $|\xi|$.

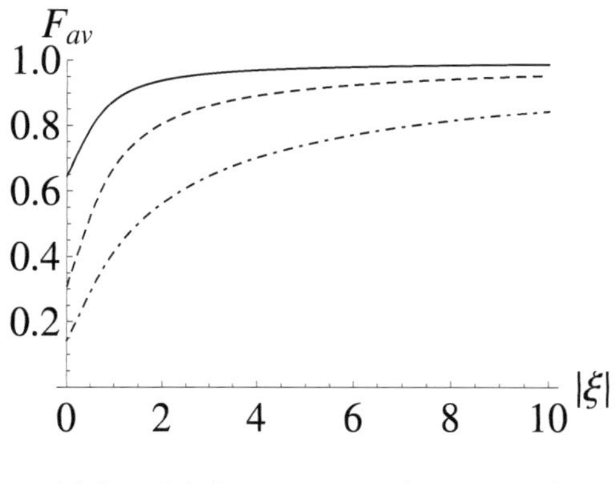

Figure 2. F_{av} as a fuction of $|\xi|$ and $|\alpha|$. From top (solid lines) to bottom (dot-dashed lines) corresponding to $|\alpha| = 0.5$, 1 and 2, respectively.

5. Conclusions

In this paper, we have investigated the quantum teleportation process via the pair coherent states by using two indicator protocols. In the measuring of the orthogonal quadrature components protocol, the average fidelity of the quantum teleportation process does not depend on the amplitude of the input state $|\alpha|$ but depends on the amplitude of the pair coherent states $|\xi|$ and q. In this protocol, F_{av} can approach to the unit if q becomes very big. That result plays an important role in the improvement of quantum teleportation fidelity comparing with the Gabris-Agarwal protocol [11]. In the measuring of the photon number sum and phase difference protocol, the average fidelity of the quantum teleportation process depends on both the amplitudes of the input state $|\alpha|$ and the pair coherent states $|\xi|$. In this protocol, F_{av} can also approach to the unit in the limit of the big amplitude value of the pair coherent states $|\xi|$. The difficulty of this protocol is the failure of the teleportation process when amplitude of teleported state becomes very big. However, those results have indicated that we can base on the relation characteristic between entanglement sources and teleported state to propose conditions for the success of the quantum teleportation process.

Acknowledgments
This research is funded by Vietnam National Foundation for Science and Technology Development (NAFOSTED) under grant number 103.01-2014.53.

References

[1] Bennett C H, Brassard G, Crepeau C, Jozsa R, Peres A and Wootters W K 1993 *Phys. Rev. Lett.* **70** 1895
[2] Furusawa A *et al* 1998 *Science* **282** 706
[3] Hofmann H F, Ide T and Kobayashi T 2000 *Phys. Rev. A* **62** 062304
[4] Janszky J, Koniorczyk M and Gabris A 2001 *Phys. Rev. A* **64** 034302
[5] Milburn G J and Braunstein S L 1999 *Phys. Rev. A* **60** 937
[6] Cochrane P T, Milburn G J and Munro W J 2000 *Phys. Rev. A* **62** 062307
[7] Karlsson A and Bourennane M 1998 *Phys. Rev. A* **58** 6
[8] An N B 2003 *Phys. Rev. A* **68** 022321
[9] Pirandola S, Eisert J, Weedbrook C, Furusawa A and Braunstein S L 2015 *Nature Photonics* **9**
[10] Opatrny T, Kurizki G and Welsch D G 2000 *Phys. Rev. A* **61** 032302
[11] Gabris A and Agarwal G S 2007 *Int. J. Quantum. Inform.* **5** 17
[12] Hou L L, Jia K H and Chen X F 2016 *Optik* **127**
[13] Li S B, Wu R K, Wang Q M and Xu J B 2004 *Phys. Lett. A* **325**
[14] Hoai N T X and Duc T M 2016 *Int. J. Mod. Phys. B* **30** 1650032
[15] Ulanov A E, Sychev D, Pushkina A A, Fedorov I A and Lvovsky A I 2017 *Phys. Rev. Lett.* **118** 160501
[16] Agarwal G S 1986 *Phys. Rev. Lett.* **57** 827
[17] Dong Y L, Zou X B and Guo G K 2008 *Phys. Lett. A* **372**
[18] Gery C C, Mimih J and Birrittella R 2011 *Phys. Rev. A* **84** 023810
[19] Duc T M, Dinh D H and Dat T Q 2016 *Int. J. Theor. Phys.* **55** 3027
[20] Kurucz Z, Adam P, Kis Z and Janszky J 2005 *Phys. Rev. A* **72** 052315
[21] Luis A and Sanchez-Soto L L 1996 *Phy. Rev. A* **51** 1

Study on elastic deformation of substitution alloy AB with interstitial atom C and BCC structure under pressure

Hoc Q Nguyen[1] and Hien D Nguyen[2]

[1] Hanoi National University of Education, 136 Xuan Thuy, Cau Giay, Hanoi, Vietnam
[2] Mac Dinh Chi High School, Chu Pah District, Gia Lai Province, Vietnam

E-mail: n.duchien@gmail.com

Abstract. The analytic expressions of the free energy, the mean nearest neighbor distance between two atoms, the elastic moduli such as the Young modulus E, the bulk modulus K, the rigidity modulus G and the elastic constants C_{11}, C_{12}, C_{44} for substitution alloy AB with interstitial atom C and BCC structure under pressure are derived from the statistical moment method. The elastic deformations of main metal A, substitution alloy AB and interstitial alloy AC are special cases of elastic deformation for alloy ABC. The theoretical results are applied to alloy FeCrSi. The numerical results for alloy FeCrSi are compared with the numerical results for main metal Fe, substitution alloy FeCr, interstitial alloy FeSi and experiments.

1. Introduction

Thermodynamic and elastic properties of interstitial alloys are specially interested by many theoretical and experimental researchers [1, 2, 5, 7-18].

The binary FeCr system is the basis for a large class of important engineering materials known as stainless steels. Stainless steels combine good corrosion resistance with attractive mechanical properties. Some other features of the FeCr alloys are also of interest, such as spin glass formation, sluggish formation of the σ phase and strong swelling resistance in an irradiative environment [11].

Iron silicides have attracted a lot of attention in recent decades, due to their unusual physical properties and functional applications [12]. The distribution of the Si-atoms in alloy FeCrSi is discussed in comparison with some common transition metal silicide structures [18].

In this paper, we build the theory of elastic deformation for substitution alloy AB with interstitial atom C and body-centered cubic (BCC) structure under pressure by the statistical moment method (SMM) [3-5].

2. Content of research

2.1. Analytic results

2.1.1. Mean nearest neighbor distance between two atoms in alloy In interstitial alloy AC with BCC structure, the cohesive energy of the atom C (in face centers of cubic unit cell) with the atoms A (in body center and peaks of cubic unit cell) in the approximation of three coordination spheres with the center C and the radii r_1, $r_1\sqrt{2}$, $r_1\sqrt{5}$ is determined by [3-5]

$$u_{0C}=\frac{1}{2}\sum_{i=1}^{n_i}\varphi_{AC}\left(r_i\right)=\frac{1}{2}\left[2\varphi_{AC}\left(r_1\right)+4\varphi_{AC}\left(r_1\sqrt{2}\right)+8\varphi_{AC}\left(r_1\sqrt{5}\right)\right]=$$
$$=\varphi_{AC}\left(r_1\right)+2\varphi_{AC}\left(r_1\sqrt{2}\right)+4\varphi_{AC}\left(r_1\sqrt{5}\right). \qquad (1)$$

where φ_{AC} is the interaction potential between the atom A and the atom C, n_i is the number of atoms on the ith coordination sphere with the radius $r_i,(i=1,2,3)$, $r_1\equiv r_{1C}=r_{01C}+y_{0A_1}(T)$ is the nearest neighbor distance between the interstitial atom C and the metallic atom A at temperature T, r_{01C} is the nearest neighbor distance between the interstitial atom C and the metallic atom A at 0K and is determined from the minimum condition of the cohesive energy u_{0C}, $y_{0A_1}(T)$ is the displacement of the atom A_1(the atom A stays in the body center of cubic unit cell) from equilibrium position at temperature T. The alloy's parameters for the atom C in the approximation of three coordination spheres have the form [3-5]

$$k_C=\frac{1}{2}\sum_i\left(\frac{\partial^2\varphi_{AC}}{\partial u_{i\beta}^2}\right)_{eq}=\varphi_{AC}^{(2)}\left(r_1\right)+\frac{\sqrt{2}}{r_1}\varphi_{AC}^{(1)}\left(r_1\sqrt{2}\right)+\frac{16}{5\sqrt{5}r_1}\varphi_{AC}^{(1)}\left(r_1\sqrt{5}\right),$$
$$\gamma_C=4\left(\gamma_{1C}+\gamma_{2C}\right),$$
$$\gamma_{1C}=\frac{1}{48}\sum_i\left(\frac{\partial^4\varphi_{AC}}{\partial u_{i\beta}^4}\right)_{eq}=\frac{1}{24}\varphi_{AC}^{(4)}(r_1)+\frac{1}{8r_1^2}\varphi_{AC}^{(2)}(r_1\sqrt{2})-\frac{\sqrt{2}}{16r_1^3}\varphi_{AC}^{(1)}(r_1)$$
$$+\frac{1}{150}\varphi_{AC}^{(4)}(r_1\sqrt{2})+\frac{4\sqrt{5}}{125r_1}\varphi_{AC}^{(3)}(r_1\sqrt{5}),$$
$$\gamma_{2C}=\frac{6}{48}\sum_i\left(\frac{\partial^4\varphi_{AC}}{\partial u_{i\alpha}^2\partial u_{i\beta}^2}\right)_{eq}=\frac{1}{4r_1}\varphi_{AC}^{(3)}(r_1)-\frac{1}{4r_1^2}\varphi_{AC}^{(2)}(r_1)+\frac{5}{8r_1^3}\varphi_{AC}^{(1)}(r_1)$$
$$+\frac{\sqrt{2}}{8r_1}\varphi_{AC}^{(3)}(r_1\sqrt{2})-\frac{1}{8r_1^2}\varphi_{AC}^{(2)}(r_{1C}\sqrt{2})+\frac{2}{25}\varphi_{AC}^{(4)}(r_1\sqrt{5})$$
$$+\frac{3}{25r_1\sqrt{5}}\varphi_{AC}^{(3)}(r_1\sqrt{5})+\frac{2}{25r_1^2}\varphi_{AC}^{(2)}(r_1\sqrt{5})-\frac{3}{25r_1^3\sqrt{5}}\varphi_{AC}^{(1)}(r_1\sqrt{5}). \qquad (2)$$

where $\varphi_{AC}^{(i)}(r_i)=\partial^2\varphi_{AC}(r_i)/\partial r_i^2(i=1,2,3,4),\alpha,\beta=x,y,z,\alpha\neq\beta$ and $u_{i\beta}$ is the displacement of the ith atom in the direction β.

The cohesive energy of the atom A_1 (which contains the interstitial atom C on the first coordination sphere) with the atoms in crystalline lattice and the corresponding alloy's parameters in the approximation of three coordination spheres with the center A_1 is determined by [3-5]

$$u_{0A_1}=u_{0A}+\varphi_{AC}\left(r_{1A_1}\right),$$
$$k_{A_1}=k_A+\frac{1}{2}\sum_i\left[\left(\frac{\partial^2\varphi_{AC}}{\partial u_{i\beta}^2}\right)_{eq}\right]_{r=r_{1A_1}}=k_A+\varphi_{AC}^{(2)}\left(r_{1A_1}\right)+\frac{5}{2r_{1A_1}}\varphi_{AC}^{(1)}\left(r_{1A_1}\right),$$
$$\gamma_{A_1}=4\left(\gamma_{1A_1}+\gamma_{2A_1}\right),$$
$$\gamma_{1A_1}=\gamma_{1A}+\frac{1}{48}\sum_i\left[\left(\frac{\partial^4\varphi_{AC}}{\partial u_{i\beta}^4}\right)_{eq}\right]_{r=r_{1A_1}}$$

$$= \gamma_{1A} + \frac{1}{24}\varphi_{AC}^{(4)}(r_{1A_1}) + \frac{1}{8r_{1A_1}^2}\varphi_{AC}^{(2)}(r_{1A_1}) + \frac{1}{8r_{1A_1}^3}\varphi_{AC}^{(1)}(r_{1A_1}),$$

$$\gamma_{2A_1} = \gamma_{2A} + \frac{6}{48}\sum_i \left[\left(\frac{\partial^4 \varphi_{AC}}{\partial u_{i\alpha}^2 \partial u_{i\beta}^2}\right)_{eq}\right]_{r=r_{1A_1}}$$

$$= \gamma_{2A} + \frac{1}{2r_{1A_1}}\varphi_{AC}^{(3)}(r_{1A_1}) - \frac{3}{4r_{1A_1}^2}\varphi_{AC}^{(2)}(r_{1A_1}) + \frac{3}{4r_{1A_1}^3}\varphi_{AC}^{(1)}(r_{1A_1}) \quad (3)$$

where $r_{1A_1} \approx r_{1C}$ is the nearest neighbor distance between the atom A_1 and atoms in crystalline lattice.

The cohesive energy of the atom A_2 (which contains the interstitial atom C on the first coordination sphere) with the atoms in crystalline lattice and the corresponding alloys parameters in the approximation of three coordination spheres with the center A_2 is determined by [3-5]

$$u_{0A_2} = u_{0A} + \varphi_{AC}\left(r_{1A_2}\right),$$

$$k_{A_2} = k_A + \frac{1}{2}\sum_i \left[\left(\frac{\partial^2 \varphi_{AC}}{\partial u_{i\beta}^2}\right)_{eq}\right]_{r=r_{1A_2}} = k_A + 2\varphi_{AC}^{(2)}\left(r_{1A_2}\right) + \frac{4}{r_{1A_2}}\varphi_{AC}^{(1)}\left(r_{1A_2}\right),$$

$$\gamma_{A_2} = 4\left(\gamma_{1A_2} + \gamma_{2A_2}\right), \gamma_{1A_2} = \gamma_{1A} + \frac{1}{48}\sum_i \left[\left(\frac{\partial^4 \varphi_{AC}}{\partial u_{i\beta}^4}\right)_{eq}\right]_{r=r_{1A_2}}$$

$$= \gamma_{1A} + \frac{1}{24}\varphi_{AC}^{(4)}(r_{1A_2}) + \frac{1}{4r_{1A_2}}\varphi_{AC}^{(3)}(r_{1A_2}) - \frac{1}{8r_{1A_2}^2}\varphi_{AC}^{(2)}(r_{1A_2}) + \frac{1}{8r_{1A_2}^3}\varphi_{AC}^{(1)}(r_{1A_2}),$$

$$\gamma_{2A_1} = \gamma_{2A} + \frac{6}{48}\sum_i \left[\left(\frac{\partial^4 \varphi_{AC}}{\partial u_{i\alpha}^2 \partial u_{i\beta}^2}\right)_{eq}\right]_{r=r_{1A_1}} = \gamma_{2A}$$

$$+ \frac{1}{8}\varphi_{AC}^{(4)}(r_{1A_2}) + \frac{1}{4r_{1A_2}}\varphi_{AC}^{(3)}(r_{1A_2}) + \frac{3}{8r_{1A_2}^2}\varphi_{AC}^{(2)}(r_{1A_2}) - \frac{3}{8r_{1A_2}^3}\varphi_{AC}^{(1)}(r_{1A_2}). \quad (4)$$

where r_{1A_1}=r_{01A_1}+$y_{0C}(T)$, r_{01A_1} is the nearest neighbor distance between the atom A_2 and atoms in crystalline lattice at 0K and is determined from the minimum condition of the cohesive energy u_{0A_2}, $y_{0C}(T)$ is the displacement of the atom C at temperature T.

In Eqs. (3) and (4), u_{0A}, k_A, γ_{1A}, γ_{2A} are the coressponding quantities in clean metal A in the approximation of two coordination sphere [3-5].

The equation of state for interstitial alloy AC with BCC structure at temperature T and pressure P is written in the form

$$Pv = -r_1\left(\frac{1}{6}\frac{\partial u_0}{\partial r_1} + \theta xcthx\frac{1}{2k}\frac{\partial k}{\partial r_1}\right). \quad (5)$$

At 0K and pressure P, this equation has the form

$$Pv = -r_1\left(\frac{\partial u_0}{\partial r_1} + \frac{\bar{h}\omega}{4k}\frac{\partial k}{\partial r_1}\right). \quad (6)$$

If knowing the form of interaction potential ϕ_{i0}, eq. (6) permits us to determine the nearest neighbor distance $r_{1X}\left(P,0\right)\left(X = C, A, A_1, A_2\right)$ at 0K and pressure P. After knowing $r_{1X}(P,0)$, we can determine alloy parametrs $k_X(P,0)$, $\gamma_{1X}(P,0)$, $\gamma_{2X}(P,0)$, $\gamma_X(P,0)$ at 0K and pressure P. After that, we can calculate the displacements [3-5]

$$\begin{aligned}
&y_{0X}(P,T) = \sqrt{\frac{2\gamma_X(P,0)\theta^2}{3k_X^3(P,0)}A_X(P,T)},\\
&A_X = a_{1X} + \sum_{i=2}^{5}\left(\frac{\gamma_X\theta}{k_X^2}\right)^i a_{iX},\\
&k_X = m\omega_X^2,\\
&x_X = \frac{\bar{h}\omega_X}{2\theta},\\
&a_{1X} = 1 + \frac{Y_X}{2},\\
&a_{2X} = \frac{13}{3} + \frac{47}{6}Y_X + \frac{23}{6}Y_X^2 + \frac{1}{2}Y_X^3,\\
&a_{3X} = -\left(\frac{25}{3} + \frac{121}{6}Y_X + \frac{50}{3}Y_X^2 + \frac{16}{3}Y_X^3 + \frac{1}{2}Y_X^4\right),\\
&a_{4X} = \frac{43}{3} + \frac{93}{2}Y_X + \frac{169}{3}Y_X^2 + \frac{83}{3}Y_X^3 + \frac{22}{4}Y_X^4 + \frac{1}{2}Y_X^5,\\
&a_{5X} = -\left(\frac{103}{3} + \frac{749}{6}Y_X + \frac{363}{3}Y_X^2 + \frac{733}{3}Y_X^3 + \frac{148}{3}Y_X^4 + \frac{53}{6}Y_X^5 + \frac{1}{2}Y_X^6\right),\\
&a_{6X} = 65 + \frac{561}{2}Y_X + \frac{1489}{3}Y_X^2 + \frac{927}{2}Y_X^3 + \frac{733}{3}Y_X^4 + \frac{145}{2}Y_X^5 + \frac{31}{3}Y_X^6 + \frac{1}{2}Y_X^7,\\
&Y_X \equiv x_X \coth x_X
\end{aligned} \tag{7}$$

From that, we derive the nearest neighbor distance $r_{1X}(P,T)$ at temperature T and pressure P

$$\begin{aligned}
&r_{1C}(P,T) = r_{1C}(P,0) + y_{A_1}(P,T), r_{1A}(P,T) = r_{1A}(P,0) + y_A(P,T),\\
&r_{1A_1}(P,T) \approx r_{1C}(P,T), r_{1A_2}(P,T) = r_{1A_2}(P,0) + y_C(P,T).
\end{aligned} \tag{8}$$

Then, we calculate the mean nearest neighbor distance in interstitial alloy AC by the expressions as follows [3-5]

$$\begin{aligned}
&\overline{r_{1A}(P,T)} = \overline{r_{1A}(P,0)} + \overline{y(P,T)}, \overline{r_{1A}(P,0)} = (1-c_C)\, r_{1A}(P,0) + c_C r'_{1A}(P,0),\\
&r'_{1A}(P,0) \approx \sqrt{3} r_{1C}(P,0), \overline{y(P,T)} = (1-7c_C)\, y_A(P,T) + c_C y_C(P,T)\\
&+2c_C y_{A_1}(P,T) + 4c_C y_{A_2}(P,T),
\end{aligned} \tag{9}$$

where $\overline{r_{1A}(P,T)}$ is the mean nearest neighbor distance between atoms A in interstitial alloy AC at pressure P and temperature T, $\overline{r_{1A}(P,0)}$ is the mean nearest neighbor distance between atoms A in interstitial alloy AC at pressure P and 0K, $r_{1A}(P,0)$ is the nearest neighbor distance between atoms A in clean metal A at pressure P and 0K, $r'_{1A}(P,0)$ is the nearest neighbor distance between atoms A in the zone containing the interstitial atom C at pressure P and 0K and c_C is the concentration of interstitial atoms C.

In alloy ABC with BCC structure (interstitial alloy AC with atoms A in peaks and body center, interstitial atom C in facer centers and then, atom B substitutes atom A in body center), the mean nearest neighbor distance between atoms A at pressure P and temperature T is determined by [3-5]

$$a_{ABC} = c_{AC} a_{AC} \frac{B_{TAC}}{B_T} + c_B a_B \frac{B_{TB}}{B_T},$$

$$
\begin{aligned}
&\overline{B_T} = c_{AC}B_{TAC} + c_B B_{TB},\\
&c_{AC} = c_A + c_C,\\
&a_{AC} = \overline{r_{1A}}(P,T),\\
&B_{TAC} = \frac{1}{\chi_{TAC}},\\
&B_{TB} = \frac{1}{\chi_{TB}}\\
&\chi_{TAC} = \frac{\left(\frac{a_{AC}}{a_{0AC}}\right)^3}{2P + \frac{\sqrt{3}}{4a_{AC}}\frac{1}{3N}\left(\frac{\partial^2\psi_{AC}}{\partial a_{AC}^2}\right)_T},\\
&\left(\frac{\partial^2\psi_{AC}}{\partial a_{AC}^2}\right)_T \approx (1-7c_C)\left(\frac{\partial^2\psi_A}{\partial a_A^2}\right)_T + c_C\left(\frac{\partial^2\psi_C}{\partial a_C^2}\right)_T + 2c_C\left(\frac{\partial^2\psi_{A_1}}{\partial a_{A_1}^2}\right)_T + 4c_C\left(\frac{\partial^2\psi_{A_2}}{\partial a_{A_2}^2}\right)_T,\\
&\frac{1}{3N}\left(\frac{\partial^2\Psi_X}{\partial a_X^2}\right)_T = \frac{1}{6}\frac{\partial^2 u_{0X}}{\partial a_X^2} + \frac{\hbar\omega_X}{4k_X}\left[\frac{\partial^2 k_X}{\partial a_X^2} - \frac{1}{2k_X}\left(\frac{\partial k_X}{\partial a_X}\right)^2\right],\\
&a_X \equiv r_{1X}(P,T).
\end{aligned}
\tag{10}
$$

The mean nearest neighbor distance between atoms A in alloy ABC at pressure P and 0K is determined by

$$
a_{0ABC} = c_{AC}a_{0AC}\frac{B_{0TAC}}{\overline{B_{0T}}} + c_B a_{0B}\frac{B_{0TB}}{\overline{B_{0T}}}. \tag{11}
$$

2.1.2. Free energy The free energy of alloy ABC with BCC structure and the condition $c_C << c_B << c_A$ has the form[5]

$$
\begin{aligned}
&\psi_{ABC} = \psi_{AC} + c_B\left(\psi_B - \psi_A\right) + TS_c^{AC} - TS_c^{ABC},\\
&\psi_{AC} = (1-7c_C)\,\psi_A + c_C\psi_C + 2c_C\psi_{A_1} + 4c_C\psi_{A_2} - TS_c^{AC},\\
&\psi_X \approx U_{0X} + \psi_{0X} + 3N\left\{\frac{\theta^2}{k_X^2}\left[\gamma_{2X}X_X^2 - \frac{2\gamma_{1X}}{3}\left(1+\frac{X_X}{2}\right)\right] + \right.\\
&\left. + \frac{2\theta^3}{k_X^4}\left[\frac{4}{3}\gamma_{2X}^2 X_X\left(1+\frac{X_X}{2}\right) - 2\left(\gamma_{1X}^2 + 2\gamma_{1X}\gamma_{2X}\right)\left(1+\frac{X_X}{2}\right)(1+X_X)\right]\right\},\\
&\psi_{0X} = 3N\theta\left[x_X + \ln(1-e^{-2x_X})\right],\\
&X_X \equiv x_X\coth x_X.
\end{aligned}
\tag{12}
$$

where ψ_X is the free energy of atom X, ψ_{AC} is the free energy of interstitial alloy AC, S_c^{AC} is the configuration entropy of interstitial alloy AC and S_c^{ABC} is the configuration entropy of alloy ABC.

2.1.3. Characteristic quantities of elastic deformation The Young modulus of alloy ABC with BCC structure at temperature T and pressure P is determined by

$$
\begin{aligned}
&E_{ABC} = c_B\left(E_B - E_A\right) + E_{AC} = c_B E_B + c_A E_A - \left(c_A + c_B\right)E_A + E_{AC} = E_{AB}\\
&-\left(c_A + c_B\right)E_A + E_{AC}, E_{AB} = c_A E_A + c_B E_B,
\end{aligned}
$$

$$E_{AC} = E_A \left(1 - 7c_C + c_C \frac{\frac{\partial^2 \psi_C}{\partial \varepsilon^2} + 2\frac{\partial^2 \psi_{A_1}}{\partial \varepsilon^2} + 4\frac{\partial^2 \psi_{A_2}}{\partial \varepsilon^2}}{\frac{\partial^2 \psi_A}{\partial \varepsilon^2}} \right),$$
$$E_A = \frac{1}{\pi . r_{1A} A_{1A}},$$
$$A_{1A} = \frac{1}{k_A} \left[1 + \frac{2\gamma_A^2 \theta^2}{k_A^4} \left(1 + \frac{1}{2} x_A cth x_A \right) (1 + x_A cth x_A) \right],$$
$$x_A = \frac{\hbar \omega_A}{2\theta},$$
$$\frac{\partial^2 \psi_X}{\partial \varepsilon^2} = \left\{ \frac{1}{2} \frac{\partial^2 U_{0X}}{\partial r_{1X}^2} + \frac{3}{4} \frac{\hbar \omega_X}{k_X} \left[\frac{\partial^2 k_X}{\partial r_{1X}^2} - \frac{1}{2k_X} \left(\frac{\partial k_X}{\partial r_{1X}} \right)^2 \right] \right\} 4r_{01X}^2$$
$$+ \left(\frac{1}{2} \frac{\partial U_{0X}}{\partial r_{1X}} + \frac{3}{2} \hbar \omega_X cth x_X \frac{1}{2k_X} \frac{\partial k_X}{\partial r_{1X}} \right) 2r_{01X},$$
$$x_X = \frac{\hbar \omega_X}{2\theta},$$
$$\omega_X = \sqrt{\frac{k_X}{m}}, \tag{13}$$

where ε is the relative deformation, E_{AB} is the Young modulus of substitution alloy AB and E_{AC} is the Young modulus of interstitial alloy AC.
The bulk modulus of alloy ABC with BCC structure at temperature T and pressure P has the form

$$K_{ABC} = \frac{E_{ABC}}{3(1 - 2\nu_{AB})}. \tag{14}$$

The rigidity modulus of alloy ABC with BCC structure at temperature T and pressure P has the form

$$G_{ABC} = \frac{E_{ABC}}{2(1 + \nu_{AB})}. \tag{15}$$

The elastic constants of alloy ABC with BCC structure at temperature T and pressure P has the form

$$C_{11ABC} = \frac{E_{ABC}(1 - \nu_{AB})}{(1 + \nu_{AB})(1 - 2\nu_{AB})}. \tag{16}$$

$$C_{12ABC} = \frac{E_{ABC}\nu_{AB}}{(1 + \nu_{AB})(1 - 2\nu_{AB})}. \tag{17}$$

$$C_{44ABC} = \frac{E_{ABC}}{2(1 + \nu_{AB})}. \tag{18}$$

The Poisson ratio of alloy ABC with BCC structure has the form

$$\nu_{ABC} = c_A \nu_A + c_B \nu_B + c_C \nu_C \approx c_A \nu_A + c_B \nu_B = \nu_{AB}. \tag{19}$$

where ν_A, ν_B and ν_A respectively are the Poisson ratioes of materials A, B and C and are determined from the experimental data.

When the concentration of interstitial atom C is equal to zero, the obtained results for alloy ABC become the coresponding results for substitution alloy AB. When the concentration of substitution atom B is equal to zero, the obtained results for alloy ABC become the coresponding results for interstitial alloy AC. When the concentrations of substitution atoms B and interstitial atoms C are equal to zero, the obtained results for alloy ABC become the coresponding results for main metal A.

2.2. Numerical results for alloy FeCrSi
For alloy FeCrSi, we use the n-m pair potential

$$\varphi(r) = \frac{D}{n-m}\left[m\left(\frac{r_0}{r}\right)^n - n\left(\frac{r_0}{r}\right)^m\right]. \tag{20}$$

where the potential parameters are given in Table 1[6]

Meterial	m	n	D $(10^{-16}erg)$	r_0 $(10^{-10}m)$
Fe	7.0	11.5	6416.448	2.4775
Cr	6.0	15.5	6612.96	2.4950
Si	6.0	12.0	45128.24	2.2950

Table 1. Potential parameters m, m, D, r_0 of materials.

Considering the interaction between Fe and Si and between Fe and Cr, we use the following approximation

$$\varphi_{Fe-Cr} \approx \frac{1}{2}\left(\varphi_{Fe-Fe} + \varphi_{Cr-Cr}\right), \varphi_{Fe-Si} \approx \frac{1}{2}\left(\varphi_{Fe-Fe} + \varphi_{Si-Si}\right). \tag{21}$$

and ignore the interaction between Cr and Si.
According to our numerical results for alloy FeCrSi at the same pressure, temperature and concentration of substitutrion atoms when the concentration of interstitial atoms increases, the mean nearest neighbor distance also increases. For example for alloy FeCrSi at T = 300K, P = 70 GPa and c_{Cr} = 10% when c_{Si} increases from 0 to 5%, r_1 increases from 2.3228 $\mathring{A}$ to 2.3989 $\mathring{A}$.

For alloy FeCrSi at the same temperature, concentration of substitution atoms and concentration of interstitial atoms when pressure increases, the mean nearest neighbor distance descreases. For example for alloy FeCrSi at T = 300K, c_{Cr} = 10%, c_{Si} = 3% when P increases from 0 to 70 GPa, r_1 descreases from 2.4715 $\mathring{A}$ to 2.3683 $\mathring{A}$.

For alloy FeCrSi at the same pressure, temperature and concentration of interstitial atoms when the concentration of substitution atoms increases, the mean nearest neighbor distance descreases. For example for alloy FeCrSi at T = 300K, P = 50 GPa, c_{Si} = 5% when c_{Cr} increases from 0 to 15% r_1 desceases from 2.4216 $\mathring{A}$ to 2.4178 $\mathring{A}$.

For alloy FeCrSi at the same pressure, concentration of substitution atoms and concentration of interstitial atoms when temperature increases, the mean nearest neighbor distance increases. For example for alloy FeCrSi at P = 0, c_{Cr} = 10% v c_{Si} = 3% when T increases from 50K to 1000K, r_1 increases from 2.4687 $\mathring{A}$ to 2.4801 $\mathring{A}$.
For alloy FeCrSi at the same pressure, temperature and concentration of substitutrion atoms when the concentration of interstitial atoms increases, the elastic moduli E, G, K increases. For example for alloy FeCrSi at T = 300K, P = 10GPa and c_{Cr} = 10% when c_{Si} increases from 0 to 5%, E increases from 18.4723.10^{10} Pa to 30.0379.10^{10} Pa.

For alloy FeCrSi at the same temperature, concentration of substitution atoms and concentration of interstitial atoms when pressure increases, the elastic moduli E, G, K increases. For example for alloy FeCrSi at T = 300K, c_{Cr} = 10%, c_{Si} = 1% when P inceases from 0 to 70GPa, E inceases from 15,2862.10^{10} Pa to 48,0400.10^{10} Pa.

For alloy FeCrSi at the same pressure, temperature and concentration of interstitial atoms when the concentration of substitution atoms increases, the elastic moduli E, G, K desceases.

For example for alloy FeCrSi at $T = 300$K, $P = 30$GPa, $c_{Si} = 5\%$ when c_{Cr} desceases from 0 to 15%, E desceases from 39.389310^{10} Pa to $39.2128.10^{10}$Pa.

For alloy FeCrSi at the same pressure, concentration of substitution atoms and concentration of interstitial atoms when temperature increases, the elastic moduli E, G, K desceases. For example for alloy FeCrSi at $P = 70$GPa, $c_{Cr} = 10\%$, $c_{Si} = 5\%$ when T increases from 50K to 1000K E desceases from $53.6239.10^{10}$Pa to $52.9651.10^{10}$Pa.

For alloy FeCrSi at the same pressure, temperature and concentration of substitutrion atoms when the concentration of interstitial atoms increases, the elastic constants, C_{11}, C_{12}, C_{44} increases. For example for alloy FeCrSi at $T = 300$K, $P = 10$GPa, $c_{Cr} = 10\%$ when c_{Si} inceases from 0 to 5%, C_{11} increases from $23.7286.10^{10}$ Pa to $38.5851.10^{10}$ Pa.

For alloy FeCrSi at the same temperature, concentration of substitution atoms and concentration of interstitial atoms when pressure increases, the elastic constants C_{11}, C_{12}, C_{44} increases. For example for alloy FeCrSi at $T = 300$K, $c_{Cr} = 10\%$, $c_{Si} = 1\%$ when P increases from 0 to 70GPa, C_{11} increases from $14.6358.10^{10}$ Pa to $61.7096.10^{10}$ Pa.

For alloy FeCrSi at the same pressure, temperature and concentration of interstitial atoms when the concentration of substitution atoms increases, the elastic constants C_{11}, C_{12}, C_{44} descreases. For example for alloy FeCrSi at $T = 300$K, $P = 30$GPa, $c_{Si} = 5\%$ when c_{Cr} increases from 0 to 15% C_{11} desceases from $51.6175.10^{10}$ Pa to $49.8943.10^{10}$ Pa.

For alloy FeCrSi at the same pressure, concentration of substitution atoms and concentration of interstitial atoms when temperature increases, the elastic constants C_{11}, C_{12}, C_{44} descreases. For example for alloy FeCrSi at $P = 70$GPa, $c_{Cr} = 10\%$, $c_{Si} = 5\%$ when T increases from 50K to 1000K, C_{11} descreases from $68.8824.10^{10}$ Pa to $68.0362.10^{10}$ Pa.

When the concentration of substitution atoms and the concentration of interstitial atoms are equal to zero, the mean nearest neighbor distance, the elastic moduli and the elastic constants of alloy FeCrSi becomes the mean nearest neighbor distance, the elastic moduli and the elastic constants of metal Fe. The dependence of mean nearest neighbor distance, the elastic moduli and the elastic constants on pressure and concentration of interstitial atoms for alloy FeCrSi is the same as the dependence of mean nearest neighbor distance, the elastic moduli and the elastic constants on pressure and concentration of interstitial atoms for interstitial alloy FeSi. The dependence of mean nearest neighbor distance, the elastic moduli and the elastic constants on pressure and concentration of substitution atoms for alloy FeCrSi is the same as the dependence of mean nearest neighbor distance, the elastic moduli and the elastic constants on pressure and concentration of substitution atoms for substitution alloy FeCr.

The dependences of elastic moduli E, K, G on temperature for Fe at $P = 0$ are given in Table 2. The dependences of nearest neighbor distance a, relative change of volume $\frac{\Delta V}{V}$, elastic moduli E, K, G and elastic constants C_{11}, C_{12}, C_{44} on pressure for Fe at $T = 300$K are given in Table 3. The nearest neighbor distance and the elastic moduli E, G of Fe at $P = 0$, $T = 300$K according to SMM and EXPT [7, 8] are given in Table 4. The dependences of elastic moduli E, K, G and elastic constants C_{11}, C_{12}, C_{44} on pressure for Fe-5%Si at $T = 300$K are given in Table 5. The dependences of elastic moduli E, K, G and elastic constants C_{11}, C_{12}, C_{44} on concentration of substitution atoms Cr for Fe-Cr at $T = 300$K and $P = 0$ are given in Table 6.

The isothermal modulus of FeSi at $T = 300$K and $P = 0$ according to the SMM, the experiments (EXPT) and the tight-binding (TB) method are given in Table 7. The lattice constant, the volume of unit cell and the bulk modulus of FeSi at $T = 300$K and $P = 0$ according to the SMM, the DFT and the EXPT are given in Tables 8, 9 and 10. Note that in these tables, FPPW is the first-principles plane-wave method combined with the quasi-harmonic Debye model, DFT^a is the DFT with ultrasoft non-norm-conserving Vanderbilt pseudopotentials in GGA, DFT^b is the DFT with Hartwigsen-Gedecker-Hutter pseudopotentials in GGA, DFT^c is the DFT with Hartwigsen-Gedecker-Hutter pseudopotentials in LDA, DFT^d is the DFT with ultrasoft pseudopotentials in LSD, DFT^e is the DFT with ultrasoft pseudopotentials in GGA.

IOP Conf. Series: Journal of Physics: Conf. Series **1034** (2018) 012005 doi:10.1088/1742-6596/1034/1/012005

$T(K)$	100	200	300	500	700	1000
$E\ (10^{10}Pa)$	22.48	21.71	20.83	18.79	16.37	12.25
$K\ (10^{10}Pa)$	15.61	15.07	14.47	13.05	11.37	6.50
$K\ (10^{10}Pa)$	8.92	8.61	8.27	7.46	6.50	4.86

Table 2. Dependences of elastic moduli E, K, G on temperature for Fe at $P = 0$.

$T(K)$	10	30	50	70
$a\ (10^{-10}m)$	2.3797	2.3486	2.3247	2.3052
$\frac{\Delta V}{V}\ (10^{-10})$	9.7	7.8	6.6	5.8
$E\ (10^{10}Pa)$	3.19	4.11	4.98	5.82
$K\ (10^{10}Pa)$	2.21	2.86	3.46	4.04
$G\ (10^{10}Pa)$	1.26	1.63	1.98	2.31
$C_{11}\ (10^{11}Pa)$	3.90	5.03	6.09	7.12
$C_{12}\ (10^{11}Pa)$	1.37	1.77	2.14	2.50
$C_{44}\ (10^{11}Pa)$	1.26	1.63	1.98	2.31

Table 3. Dependencse of nearest neighbor distancet a, relative change of volume $\frac{\Delta V}{V}$, elastic moduli E, K, G and elastic constants C_{11}, C_{12}, C_{44} on pressure for Fe at $T = 300$K.

Method	a (Å)	E $(10^{10}Pa)$	G $(10^{10}Pa)$	C_{11} $(10^{11}Pa)$	C_{12} $(10^{11}Pa)$	C_{44} $(10^{11}Pa)$
SMM	2.4298	20.83	8.27	2.48	1.01	0.73
EXPT	2.3729[7]	20.98[8]	8.12[8]	2.33	1.35[8]	1.18[8]

Table 4. Nearest neighbor distance, elastic moduli E, G and elastic constants C_{11}, C_{12}, C_{44} of Fe at $P = 0$, $T = 300$K according to SMM and EXPT [7, 8].

$P(GPa)$	10	30	50	70
$E\ (10^{10}Pa)$	26.19	35.16	43.56	51.58
$K\ (10^{10}Pa)$	18.19	24.42	30.25	35.82
$G\ (10^{10}Pa)$	10.39	13.93	17.29	20.47
$C_{11}\ (10^{10}Pa)$	32.04	43.02	53.30	63.11
$C_{12}\ (10^{10}Pa)$	11.26	15.12	18.73	22.17
$C_{44}\ (10^{10}Pa)$	10.39	13.93	17.29	20.47

Table 5. Dependences of elastic moduli E, K, G and elastic constants C_{11}, C_{12}, C_{44} on pressure for Fe-5%Si at T = 300K.

$c_{Cr}(\%)$	3	6	9	12	15
$E\ (10^{10}Pa)$	20.91	20.98	21.06	21.14	21.14
$K\ (10^{10}Pa)$	14.65	14.83	15.02	15.21	15.40
$G\ (10^{10}Pa)$	8.28	8.30	8.32	8.33	8.35
$C_{11}(10^{10}Pa)$	25.69	25.90	26.11	26.32	26.54
$C_{12}(10^{10}Pa)$	9.12	9.30	9.48	9.66	9.84
$C_{44}(10^{10}Pa)$	8.28	8.30	8.32	8.33	8.35

Table 6. Dependences of elastic moduli E, K, G and elastic constants C_{11}, C_{12}, C_{44} on concentration of substitution atoms Cr for Fe-Cr at T = 300K and P=0.

Method	SMM	EXPT[9]	EXPT[10]
B_T[GPa]	170.09	168	281

Table 7. Isothermal elastic modulus of Fe at T = 300K and P = 0.

Method	SMM	FPPW +QHD [12]	DFT^a [13]	DFT^b [14]	DFT^c [14]	DFT^d [15]	DFT^e [15]	EXPT [16]	EXPT [17]
$\frac{2a}{\sqrt{3}}$ (Å)	2.890	2.785	2.768	2.772	2.708	2.72	2.77	2.791	2.773

Table 8. Lattice constant of FeSi at T = 300K and P = 0.

Method	SMM	FPPW+QHD [12]	EXPT[16]	EXPT[17]
$V = (\frac{2a}{\sqrt{3}})^3$, Å^3	24.14	21.6	21.74 ± 0.02	21.32

Table 9. Volume of unit cell of FeSi at T = 300K and P = 0.

IOP Conf. Series: Journal of Physics: Conf. Series **1034** (2018) 012005 doi:10.1088/1742-6596/1034/1/012005

Method	SMM	FPPW +QHD [12]	DFT^a [13]	DFT^d [15]	DFT^e [15]	EXPT [16]	EXPT [17]
K(GPa)	194.395	199	226	263	221	184 ± 5	225

Table 10. Bulk modulus of FeSi at $T = 300$K and $P = 0$.

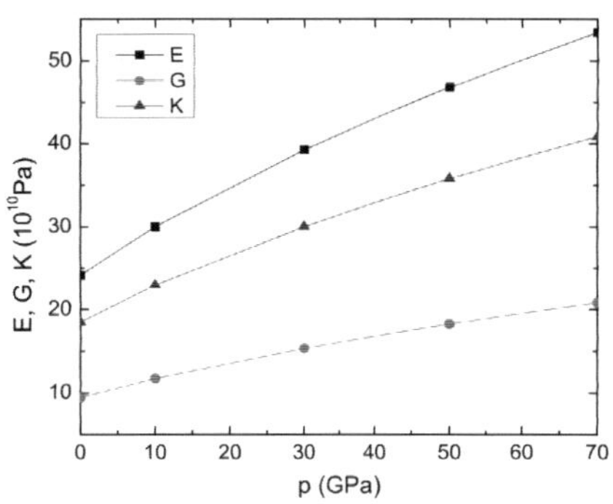

Figure 1. Dependence of elastic moduli E, G, K on pressure for alloy Fe-10%Cr-5%Si at $T = 300$K.

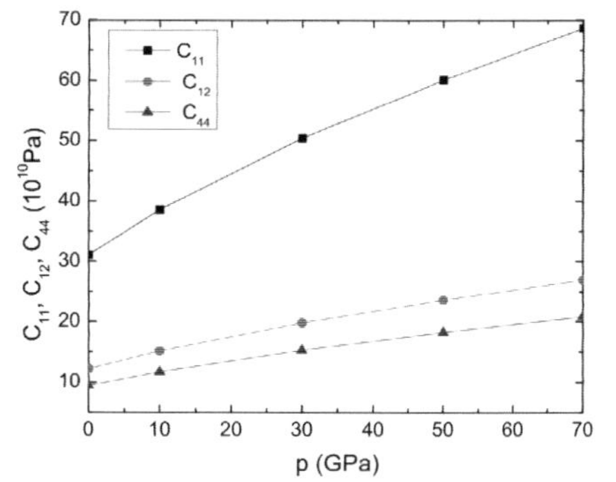

Figure 2. Dependence of elastic constants C_{11}, C_{12}, C_{44} on pressure for alloy Fe-10%Cr-5%Si at T = 300K.

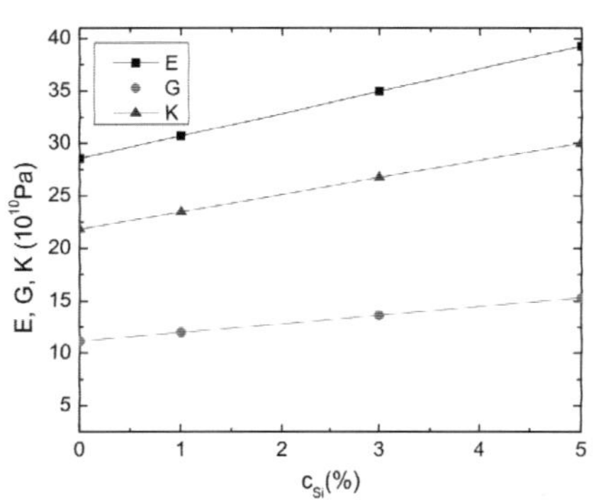

Figure 3. Dependence of elastic moduli E, G, K on concentration of Si for alloy Fe-10%Cr-xSi at $P = 30$GPa and $T = 300$K.

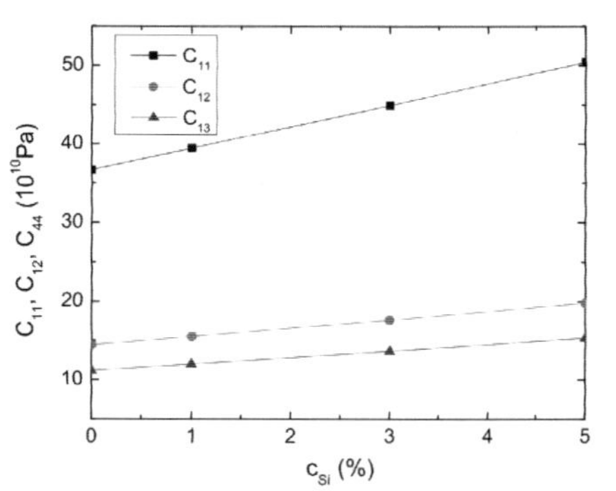

Figure 4. Dependence of elastic constants C_{11}, C_{12}, C_{44} on concentration of Si for alloy Fe-10%Cr-xSi at $P =$ 30GPa and $T = 300$K.

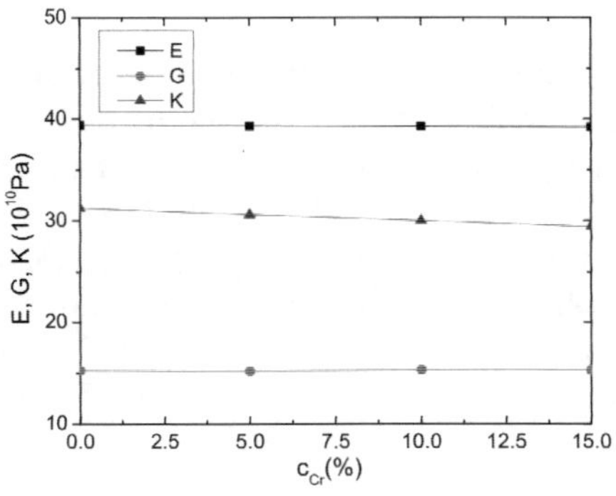

Figure 5. Dependence of elastic moduli E, G, K on concentration of Cr for alloy Fe-xCr-5%Si at $P = 30$GPa and $T = 300$K.

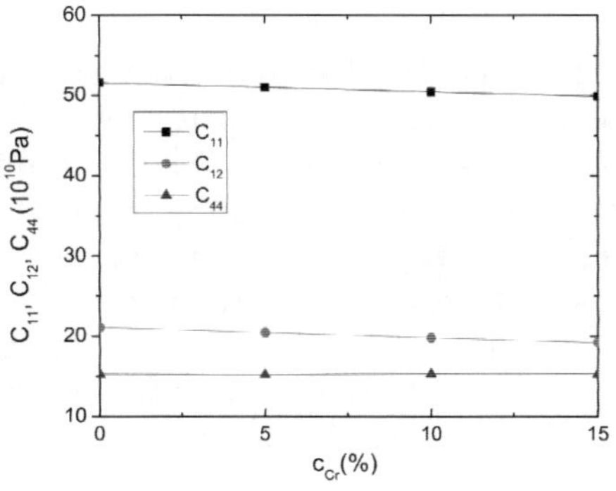

Figure 6. Dependence of elastic constants C_{11}, C_{12}, C_{44} on concentration of Cr for alloy Fe-xCr-5%Si at $P = 30$GPa and $T = 300$K.

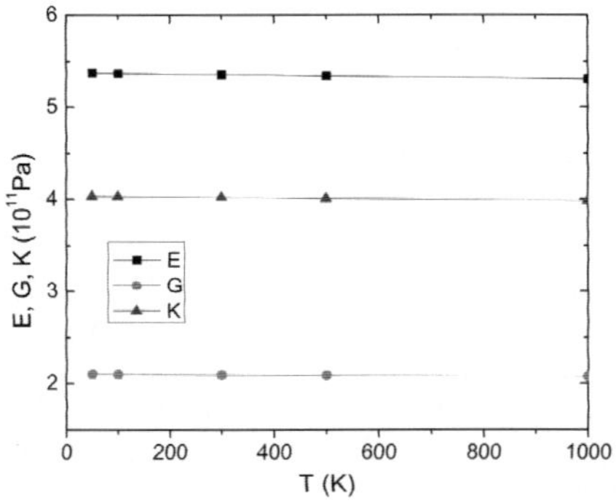

Figure 7. Dependence of elastic moduli E, G, K on temperature for alloy Fe-15%Cr-5%Si at $P = 70$GPa.

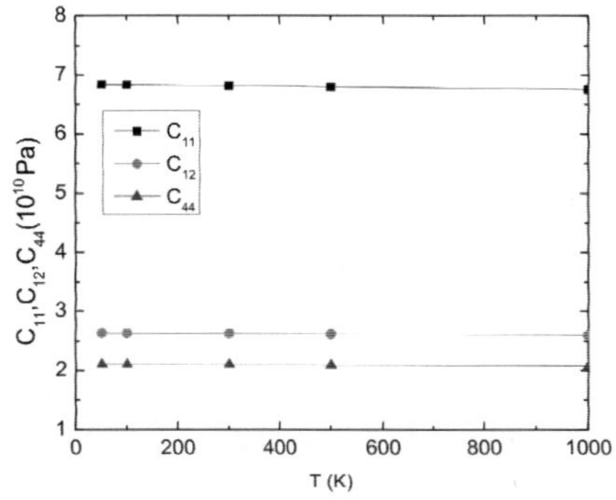

Figure 8. Dependence of elastic constants C_{11}, C_{12}, C_{44} on temperature for alloy Fe-15%xCr-5%Si at $P = 70$GPa.

IOP Conf. Series: Journal of Physics: Conf. Series **1034** (2018) 012005 doi:10.1088/1742-6596/1034/1/012005

3. Conclustion

The analytic expressions of the free energy, the mean nearest neighbor distance between two atoms, the elastic moduli such as the Young modulus, the bulk modulus, the rigidity modulus and the elastic constants depending on temperature, concentration of substitution atoms and concentration of interstitial atoms for substitution alloy AB with interstitial atom C and BCC structure under pressure are derived by the SMM. The numerical results for alloy FeCrSi are in good agreement with the numerical results for substitution alloy FeCr, interstitial alloy FeSi and main metal Fe. Temperature changes from 5 to 1000K, pressure changes from 0 to 70 GPa, the concentration of substitution atoms Cr changes from 0 to 15% and the concentration of interstitial atoms Si changes from 0 to 5%.

References

[1] Mironov K E, *Interstitial alloy.* Plenum Press, New York, 1967
[2] Smirnov A A, *Theory of Interstitial Alloys*, Nauka, Moscow, 1979 (in Russian)
[3] Tang N and Hung V V, *Phys. Stat. Sol. (b)* **149**(1988)511; **161**(1990)165; **162**(1990)371; **162**(1990) 379
[4] Hung V V, *Statistical moment method in studying thwermodynamic and elastic property of crystal*, HNUE Publishing House, 2009
[5] Hoc N Q, Vinh D Q, Tinh B D, Loan T T C, Phuong N L, Hue T T and Thuy D T T, Thermodynamic properties of binary interstitial alloys with a BCC structure: dependence on temperature and concentration of interstitial atoms, *Journal of Science of HNUE, Math. and Phys. Sci.* **60**, 7 (2015) pp.146-155
[6] Magomedov M N, *J. Fiz. Khimic* **61**(1987)1003 (in Russian)
[7] Lide D R, *CRC Handbook of Chemistry and Physics*, 86th Ed., Taylor & Francis, Boca Raton London, New York, Singapore, 2005.
[8] Tikhonov L V et al., *Mechanical properties of metals and alloys*, Kiev, 1986 (in Russian)
[9] Hyunchae Cyunn and Choong-Shik Yoo, Phys. Rev., B **59** (1999)8526
[10] Mehl M J, Papaconstantopoulos D A, *Phys. Rev.*, B **54** (1996)4519.
[11] Xiong W, Selleby M, Chen Q, Odqvist J and Du Y, Phase equilibria and thermodynamic properties in the FeCr system, *Critical Review in Solid State and Materials Sciences* **35** (2010)125
[12] Zhao K M, Jiang G and Wang L, Electronic and thermodynamic properties of B2-FeSi from first principles, Physica B **406**(2011)363
[13] Vocadlo L, Price G D and Wood I G, *Acta Crystalogr.Sect.* B**55**(1999)484
[14] Caracas R, Wentzcovich R, *Geophys.Res.Lett.***31**(2004)L20603
[15] Moroni E G, Wolf W, Hafner I and Podloucky R, *Phys.Rev.*B59(1999)860
[16] Dobson D P, Crichton W A, Bouvier P P, Vocadlo L and Wood I G, *Geophys. Res. Lett.* **30**(2003)1
[17] Ono S, Kikegawa T and Ohishi Y, *Eur.J.Mineral* **19**(2007)183
[18] Verkina B I and Prikhotko A Ph, *Kriokristallu*, Kiev (1983)(in Russian)
[19] Aronsson B and Lundstrom T, *Acta Chemica Scandinavica* **11**(1957)365

IOP Conf. Series: Journal of Physics: Conf. Series **1034** (2018) 012006 doi:10.1088/1742-6596/1034/1/012006

Drude-jellium model for the microwave conductivity of electrolyte solutions

Tran Thi Nhan[1], Luong Thi Theu[2], Le Tuan[3], and Nguyen Ai Viet[4]

[1] Hanoi University of Industry, 298 Cau Dien Road, Hanoi, Vietnam

[2] Hanoi Pedagogical University Number 2, Xuan Hoa, Vinh Phuc, Vietnam

[3] Hanoi University of Science and Technology, 1 Dai Co Viet Road, Hanoi, Vietnam

[4] Institute of Physics, Vietnam Academy of Science and Technology, 10 Dao Tan, Hanoi, Vietnam

E-mail: tran.nhan@haui.edu.vn

Abstract. The microwave conductivity characteristics of electrolyte solutions have attracted much interest of researchers because a good understanding of their properties plays a key role to study fundamental processes in biology and chemistry. In this work, we consider the solution of sodium chloride as a plasma consisting of ions with water background. Its plasmon frequency is calculated by the jellium theory. The linear dependence of the microwave conductivity on the ion concentration of the electrolyte solutions is explained by a microscopic approach and described by a combination of this plasmon relationship and the simplified Drude formula for dielectric constant. Furthermore, the dependence of the microwave conductivity on the frequency of the salt solution is also examined. We suggest that it obeys the logistic distribution. We found a good agreement between theoretical calculations and experimental data. The values of the damping coefficient γ for the conductive solutions at low frequencies and the cutting frequency are estimated. The linear dependence of the diffusion coefficient on the temperature of the salt solution is also shown, in similarity with the result in the other model. The application of the Drude-jellium model could be done for the other electrolyte solutions in order to study theirs electro-dynamic properties.

1. Introduction

The important role of aqueous electrolyte solutions in biological systems is widely recognized. Researches on their characteristics could offer a key to investigate electrokinetic phenomena which happen in living cells, including charge transfer, electro-osmosis, electrophoresis [1] as well as electrochemical processes [2]. It is noticeable that the solvation of ions in aqueous media is a fundamental process in biology and chemistry [2–5]. A good understanding of the conductivity of aqueous electrolyte solutions at microwave frequencies could give a powerful way to investigate the dynamical processes in solution media [2, 6–8]. Researches on the microwave conductivity of electrolyte solution also help us have a deeper comprehension of the interaction mechanisms between electromagnetic waves and biological tissues surrounded by an aqueous buffer containing ions with different concentrations [9].

Electrical properties of electrolyte solution have been widely interested over the last 120 years [4]. Recently, the microwave conductivity of the salt solution was determined by measuring its dielectric constant via the combination of the Debye and Drude models [10]. The linear increase of the microwave conductivity of the sodium chloride solution at room temperature versus its ion density was observed. This dependence was explained by the simplified Drude model. It is interesting that the microwave conductivity of the sodium chloride solution holds constant at low frequency (under 8 GHz) but it was higher than that in the static regime [11,12]. In the opposite limit, the microwave conductivity of the solution obviously decreases as the frequency increases and reaches zero at high enough frequencies. However, the science behind the microwave conductivity behavior of the salt solution is poorly understood due to the absence of theoretical approaches. So, a further research to provide a microscopic model is needed for a good understanding of microwave conductivity behavior for the sodium chloride solution as well as the other electrolyte solutions.

In this paper, considering the electrolyte solution of sodium chloride as a plasma with water background, its plasmon frequency is calculated by the jellium theory. The concentration and frequency dependence of the microwave conductivity of the salt solution is described by a simple approach called the Drude-jellium model. The value of the damping constant of this solution in the regime of low frequency is estimated. The comparison between theoretical calculations and empirical data is performed to validate the Drude-jellium model. The dependence of the diffusion coefficient on the temperature of the salt solution at low frequencies is also considered by our model.

2. The jellium theory for electrolyte solution

2.1. Jellium theory

Let us consider the motion of one species of charged particles with density N, charged e (e is the electron charge), and mass m in a neutralizing rigid continuous background of positive charge. If their mass is large, these charged particles have a classical dynamics. So, the motion of particles is affected by the Coulomb potentials $\varphi = e^2/4\pi\varepsilon_0 r$, where ε_0 is the electric constant and r denotes the distance between two charged particles. The jellium theory [12, 13] may be a useful tool to investigate dynamic processes which take place in the system. The Coulomb interaction in the jellium theory can be written as

$$U = \frac{1}{2}\int d\mathbf{r}d\mathbf{r}'\varphi(\mathbf{r}-\mathbf{r}')\delta n(\mathbf{r})\delta n(\mathbf{r}'), \tag{1}$$

where $\delta n(\mathbf{r})$ is a small disturbances of density at $\mathbf{r}$. Let's employ now Fourier representation for $\delta n(\mathbf{r})$ in its spatial Fourier coefficients at wavevector $\mathbf{q}$

$$\delta n(\mathbf{r}) = \frac{1}{\sqrt{N_0}}\sum_{\mathbf{q}}\delta n(\mathbf{q})e^{i\mathbf{qr}}, \tag{2}$$

where $N_0 = NV$ is the total number of charged particles containing in volume V and $\delta n(\mathbf{q}) = \frac{N}{\sqrt{N_0}}\int d\mathbf{r}\delta n(\mathbf{r})e^{-i\mathbf{qr}}$. Likewise, we also have

$$\varphi(\mathbf{r}) = \frac{1}{\sqrt{V}}\sum_{\mathbf{q}}\varphi(\mathbf{q})e^{i\mathbf{qr}}, \tag{3}$$

where $\varphi(\mathbf{q}) = \int d\mathbf{r}\varphi(\mathbf{r})e^{-i\mathbf{qr}}$. It is noticeable that $\varphi(q) = e^2/4\pi\varepsilon_0 q^2$ is the Fourier representation of the Coulomb potential. Substituting (2) and (3) into (1), the Coulomb interaction for the system becomes

$$U = \frac{1}{2N}\sum_{\mathbf{q}}\varphi(q)\delta n(\mathbf{q})\delta n(-\mathbf{q}). \tag{4}$$

It is possible to represent the small variation in the density $\delta n = -N div\mathbf{u}$, where $\mathbf{u}$ is the vector of displacement. This representation holds for the condition $\mathbf{qu}(\mathbf{r}) \ll 1$. On the other hand, the Coulomb interaction involves only longitudinal part of the displacement vector $\mathbf{u}(\mathbf{q})$ along the wavevector $\mathbf{q}$, we may have $\mathbf{u}(\mathbf{q}) = (\mathbf{q}/q)u(\mathbf{q})$. Thus, we can write $\delta n(\mathbf{q}) = -iN\mathbf{q}\mathbf{u}(\mathbf{q})$ with $\delta n^*(-\mathbf{q}) = \delta n(\mathbf{q})$, $\mathbf{u}^*(-\mathbf{q}) = \mathbf{u}(\mathbf{q})$, and $u^*(-\mathbf{q}) = -u(\mathbf{q})$. The Coulomb interaction according to the relation (4) is given by

$$U = -\frac{N}{2}\sum_{\mathbf{q}} q^2\varphi(q)u(\mathbf{q})u(-\mathbf{q}). \tag{5}$$

Furthermore, the kinetic energy in the displacement wavevector representation for the system is defined by

$$T = \frac{1}{2}\int d\mathbf{r}Nm\dot{\mathbf{u}}^2 = -\frac{1}{2}m\sum_{\mathbf{q}} \dot{u}(\mathbf{q})\dot{u}(-\mathbf{q}). \tag{6}$$

So, the equations of motion for the system derived from the Lagrange function are

$$m\ddot{u}(\mathbf{q}) + Nq^2\varphi(q)u(\mathbf{q}) = 0, \tag{7}$$

which gives the well-known plasma frequency $\omega_0^2 = e^2N/\varepsilon_0 m$.

2.2. Plasma oscillation for the salt solution

We consider that the salt solution (NaCl) is a plasma consisting of two ionic species, cation Na^+ and anion Cl^- with the density of each ion being N. The mass of cation Na^+ and anion Cl^- are denoted by m_1 and m_2, respectively. In this situation, we consider both ionic particles continuously distributed in a neutral continuous background of pure liquid water. Because of their large mass, the ions have a classical dynamics. We can apply the jellium model to describe the collective oscillation for the electrolyte solution of NaCl. Here, the equations of ionic motion can be derived by writing two equations, one for the displacement vector $\mathbf{u}_1$ of the cation Na^+ and another one for the displacement vector $\mathbf{u}_2$ of the anion Cl^-. In similarity with the previous case, we limit ourselves to consider the motion of ions in the water under the action of the Coulomb potentials $\varphi_{Na-Na} = \varphi(q)$, $\varphi_{Na-Cl} = -\varphi(q)$, and $\varphi_{Cl-Cl} = \varphi(q)$, where $\varphi(q) = e^2/4\pi\varepsilon_0 q^2$. As a result, the Coulomb interactions for the solution are defined by

$$\begin{aligned} U_{Na-Na} &= -\tfrac{N}{2}\textstyle\sum_{\mathbf{q}} q^2\varphi(q)u_1(\mathbf{q})u_1(-\mathbf{q}), \\ U_{Cl-Cl} &= -\tfrac{N}{2}\textstyle\sum_{\mathbf{q}} q^2\varphi(q)u_2(\mathbf{q})u_2(-\mathbf{q}), \\ U_{Na-Cl} &= N\textstyle\sum_{\mathbf{q}} q^2\varphi(q)u_1(\mathbf{q})u_2(-\mathbf{q}). \end{aligned} \tag{8}$$

The kinetic energy T_s of the system is represented by

$$T_s = -\frac{1}{2}[m_1\sum_{\mathbf{q}} \dot{u}_1(\mathbf{q})\dot{u}_1(-\mathbf{q}) + m_2\sum_{\mathbf{q}} \dot{u}_2(\mathbf{q})\dot{u}_2(-\mathbf{q})]. \tag{9}$$

Consequently, the motions of both ions are written as

$$\begin{aligned} m_1\ddot{u}_1(\mathbf{q}) + Nq^2\varphi(q)u_1(\mathbf{q}) - 2Nq^2\varphi(q)u_2(\mathbf{q}) = 0, \\ m_2\ddot{u}_2(\mathbf{q}) + Nq^2\varphi(q)u_2(\mathbf{q}) - 2Nq^2\varphi(q)u_1(\mathbf{q}) = 0. \end{aligned} \tag{10}$$

The solution of these equations in the long wavelength limit $\mathbf{q} \to 0$ is

$$\omega_p^2 = \frac{Ne^2}{\varepsilon_0}(\frac{1}{m_1} + \frac{1}{m_2}), \tag{11}$$

which is considered as the plasmon frequency of the salt solution. According to this relation, for example, the plasmon frequency for the solution of sodium chloride with concentration 6.93% is approximately 10^{12} Hz.

IOP Conf. Series: Journal of Physics: Conf. Series **1034** (2018) 012006 doi:10.1088/1742-6596/1034/1/012006

It is possible to generalize this model to a system of multi-component plasma consisting of several ions labeled by i, each ionic species with the density N_i, charge $z_i e$ (z_i is the reduced effective electron charge), and mass m_i. In similarity with the previous case, the plasmon frequency of such a system could be obtained in the following form [13]

$$\omega_p^2 = \sum_i \frac{N_i z_i^2 e^2}{\varepsilon_0 m_i}. \tag{12}$$

3. Microwave conductivity of electrolyte solution in the Drude-jellium model

It is well-known that the relative permittivity ε_D of metallic conductors obeys the Drude dispersion formula with the real part $\varepsilon_D^{'}$ and the imaginary part $\varepsilon_D^{''}$

$$\varepsilon_D(\omega) = \varepsilon_D^{'} + j\varepsilon_D^{''} = (1 - \frac{\omega_0^2}{\omega^2 + \gamma_e^2}) + j\frac{\omega_0^2}{\omega}\frac{\gamma_e}{\omega^2 + \gamma_e^2}, \tag{13}$$

where ω_0 is the plasmon frequency of the system, γ_e is the damping coefficient relating to the mean distance between subsequent collisions of an electron, and ω is the frequency. We note that the damping coefficient can be written by $\gamma_e = 1/\tau_e$, where τ_e is the relaxation time according to the average time between subsequent collisions of an electron. For electromagnetic waves whose frequency is satisfied with the condition $\omega \ll \gamma_e$, the Drude model for the relative permittivity of metal becomes simpler with the pure imaginary part, $\varepsilon(\omega) = \varepsilon^{''}(\omega) = j\sigma^0/(\varepsilon_0\omega)$, where $\sigma^0 = Ne^2/(m\gamma_e) = \omega_p^2\varepsilon_0/\gamma_e$ is the static conductivity of metal.

In similarity with the electron gas in the metal, the ions in the solution of sodium chloride are responsible for the conductivity with the water background. We assume that the interaction between ions and water molecules is negligible. Since the plasmon frequency of solution determined by (11), we have

$$\sigma_{solu}^0 = \frac{Ne^2}{\gamma_0}(\frac{1}{m_1} + \frac{1}{m_2}), \tag{14}$$

where σ_{solu}^0 is called the static conductivity for the sodium chloride in the Drude-jellium model. In accordance with the relation (14), the static conductivity of solution is linearly proportional to the density of ion N if the temperature is held constant. In fact, the concentration dependence plot for the static conductivity of the solution is a pure linearity [10] (the dashing line in the Fig. 1). It means that the damping constant γ_0 of the solution in the static regime is a constant at a definite temperature. According to the relation (14), its value can be estimated by determining the slop of the concentration dependence plot for the static conductivity of the solution, $\gamma_0 \approx 10^{14}\ s^{-1}$ [10], approximately the damping constant of metallic conductors [14].

It is well-known that the relation between the density of ion flow J and the current density amplitude I for the electrolyte solution of sodium chloride is given by $I = eJ$. The density of ion flow under the action of an external force could be written by the following expression [15]

$$J = J_D + J_\sigma, \tag{15}$$

where J_D and J_σ are the densities of the diffusion and convective flows, respectively. In addition, the diffusion flow counteracts the convective one. Thus, the relaxation time of solution is reduced in the presence of the diffusion flow. As an external field with microwave frequency is applied, the diffusion modes are too slow to follow the external field, resulting in the absence of the diffusion flow in the density of ion flow J. Only the convective flow survives, so the relaxation time of solution at low frequencies is higher than that in the static regime. We suggest that the damping constant of electrolyte solution at low frequency is invariable denoted γ, and, we have $\gamma < \gamma_0$. In fact, the motion or relaxation of most ions can respond effectively to the microwave field at low

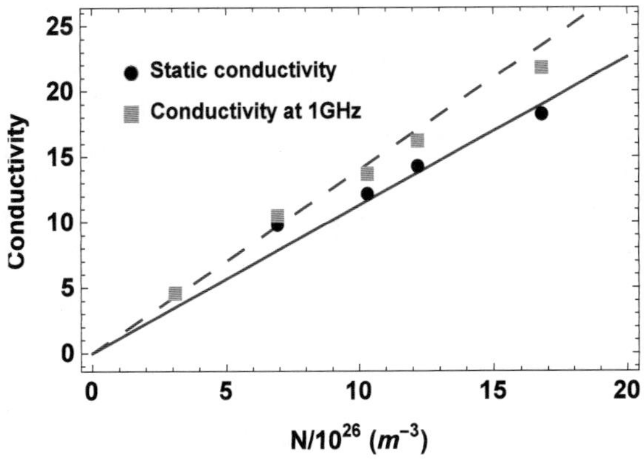

Figure 1. The AC conductivity σ^0_{max} at 1GHz of the sodium chloride solution in the Drude-jellium model versus the number of ions is shown by the solid line. The dashing line illustrates the concentration dependence of the static conductivity σ^0_{solu} for the salt solution in Ref. 10. The experimental data [10] (symbols) are also represented to validate our Drude-jellium model.

frequencies. As a result, the microwave conductivity of the sodium chloride solution is invariable at low frequency and a definite temperature, denoted σ^0_{max}. Due to $\gamma < \gamma_0$, the microwave conductivity of solution at low frequencies (below 8 GHz) is higher than its static conductivity. This behavior has ever been observed by experiments [10,11]. Thus, the microwave conductivity of solution at low frequencies is defined by

$$\sigma^0_{max} = \frac{Ne^2}{\gamma}(\frac{1}{m_1} + \frac{1}{m_2}). \tag{16}$$

In accordance with the formula (16), the microwave conductivity for the salt solution is independent of the frequency, but it linearly depends on the ionic concentration (the solid line in Fig. 1), resembling the behavior of the static conductivity mentioned above. In a similar way to derive the damping constant γ_0, the damping constant $\gamma \approx 0.78 \times 10^{14}$ s$^{-1} \approx 0.8\gamma_0$ is obtained for the salt solution at low frequencies. In the microwave regime, it is clear that the value of the damping coefficient is much higher than that of the frequency ω. So, it proves the applicability of the Drude model in the simple form.

In addition, the response of ions to the alternative external electric field becomes more and more difficult with the increase in frequency due to their large mass. Consequently, an obvious decrease of the AC conductivity of solution to zero at high enough frequencies is observed. We suggest that ω_C is the maximum frequency at which most ions can be still responsible for the conductivity. Above this frequency, the ions are responsible for the conductivity function under the action of thermal fluctuation. It is clear to see that the number of ions being responsible for the conductivity of the sodium chloride solution is satisfied with the logistic function [16] which is used in a range of fields, including biology (especially ecology) [17,18], physics (Fermi–Dirac statistics) [19], economics [20], linguistics [21], and statistics with the sigmoid's midpoint ω_C and the variable ω. In our opinion, the role of ω_C is similar to that of the cutoff frequency of physics and electrical engineering. According to the experimental data, the value of ω_C is approximately 12 GHz. As a result, the real part of the microwave conductivity $\sigma_{solu}(\omega)$ for the sodium chloride solution at room temperature $T_0 = 293$ K also obeys the logistic statistics with the curve's maximum value σ^0_{max}

$$\sigma_{solu}(\omega) = \frac{Ne^2}{\gamma}(\frac{1}{m_1} + \frac{1}{m_2})\frac{1}{1 + exp[\alpha(\omega - \omega_C)]}, \tag{17}$$

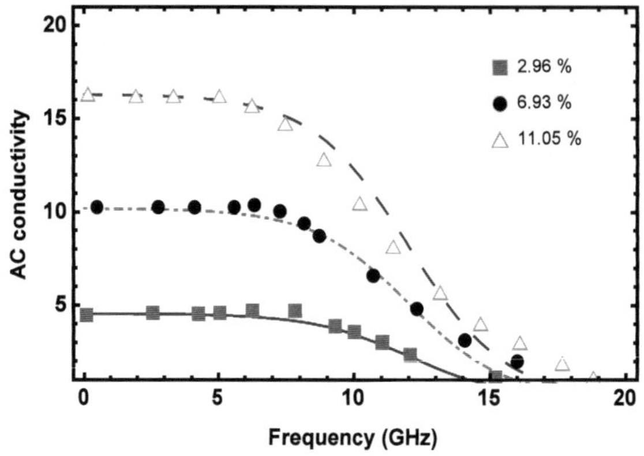

h

Figure 2. Frequency spectra of the microwave conductivity in the Drude-jellium model for the solution of sodium chloride with various concentrations 2.96 % , 6.93 % , and 11.05 % is represented by the solid line, dot-dashed line, and dashing line, respectively. A good agreement between calculations made by formula (17) and experiment data [10] (symbols) is also shown.

where $\alpha = 8.38h/k_B T_0$ (k_B is the Boltzmann constant and h is the Planck's constant) is corresponding to the steepness of the curve. In accordance with this relation, the microwave conductivity of the salt solution is invariable at low frequencies and it strongly decreases to zero at high enough frequencies. The linear concentration dependence of the microwave conductivity is also displayed. Furthermore, we found a good agreement between calculation results and experimental data in Ref. [10] (see Fig. 2). It is easily seen that the relation (17) in the Drude-jellium model describes well the frequency dependence of the microwave conductivity of the sodium chloride solution in a wide range of concentrations.

It is noticeable that the value of the cutoff frequency is much smaller than that of the plasmon frequency ($\omega_C = 10^{-2}\omega_p$). However, it is widely accepted that the value of the cutoff frequency is approximately equal to that of the plasmon frequency for the electron gas. In our opinion, it is the water background that has an impact on the ion motion resulting in the reduction in the value of the cutoff frequency in this case in comparison with the situation of the electron gas. Furthermore, it is the influence of the water background on the ion motion that also leads to a such value of the steepness of the logistic curve. Therefore, it is necessary to take into account the role of the water background in the jellium model for the electrolyte solution instead of ignoring its influence.

4. The diffusion coefficient of sodium chloride solution at low frequencies

It is well-known that the convection flow of ions in the salt solution looks like

$$J_\sigma = NbF, \tag{18}$$

where $F = eE$ is the force caused by the alternative field E acting on ions, N is the ion density and b is the ion mobility. The ion mobility combines with the diffusion coefficient D_0 by the Einstein relation $bk_BT = D_0$, where k_B is the Boltzmann constant and T is the absolute temperature. We note that the diffusion coefficient is related to the ion concentration by the Fick's first law $\mathbf{J}_D = -D_0 gradN$, where $\mathbf{J}_D$ denotes the diffusion flux vector.

As an external field with microwave frequency is applied, only the convective flow survives due to the absence of the diffusion flow in the flow density of ions J. So the current density amplitude in the electrolyte solution is given by

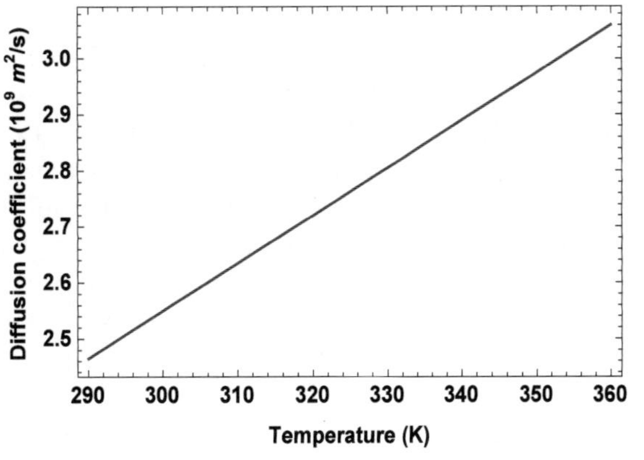

h

Figure 3. The dependence of the diffusion coefficient on the temperature of the sodium chloride solution at low frequencies is exhibited by the Drude-jellium model.

$$I = \frac{e^2 N D_0}{k_B T} E. \tag{19}$$

Thus, the microwave conductivity of the sodium chloride solution is read by

$$\sigma^0_{max}(\omega) = \frac{e^2 N D_0}{k_B T}. \tag{20}$$

In comparison between (16) and (20), the diffusion coefficient at low frequency for the solution of sodium chloride is given by

$$D_0 = (\frac{1}{m_1} + \frac{1}{m_2})\frac{k_B T}{\gamma}. \tag{21}$$

According to the relation (21), the diffusion coefficient of sodium chloride solution is a linear function of temperature. It is also independent of the solution concentration. This result is similar to that observed in experiment [11] and in agreement with the Stokes–Einstein equation. This is also an evidence to validate the Drude-jellium model for microwave conductivity of electrolyte solution suggested in the above section again.

5. Conclusion

In this work, we provided the jellium model for the electrolyte solution of sodium chloride, which was considered as a plasma consisting of ions with the water background. Its plasmon frequency was defined by our model. The linear dependence of the microwave conductivity on the ion concentration of the sodium chloride solution was pointed out by the combination of its plasmon frequency and the Drude formula in the simple form of dielectric constant. The decrease in the damping constant leading to the rise in the microwave conductivity of the solution at low frequencies in comparison with that in the static regime was interpreted due to the absence of the diffusion flow in the solution. In addition, the dependence of the microwave conductivity on the frequency of the salt solution was mentioned and explained by a microscopic approach. It obeys a famous distribution called the logistic statistics. According to our model, the microwave conductivity of the salt solution is almost dispersionless at low frequencies and it strongly decreases to zero at high enough frequencies. We found a good agreement between theoretical results and empirical data in a wide range of concentration. Furthermore, the linear dependence of the diffusion coefficient on the temperature of the solution at low frequencies being similar to the famous Stokes–Einstein was also pointed out by our model. In addition, the value of the cutoff frequency is given, but it is much lower than that of the plasmon frequency due to the influence of the water background. However, the role of the water background has not been taken into account in our jellium model. This problem also leads to the poor understanding of the physical meaning of the constant according to the steepness of the logistic curve in this work. It is necessary to modify our jellium model to provide a new jellium model in which the role of the water background is taken into consideration.

Our research on the microwave conductivity of sodium chloride solution under ambient temperature is essential for an accurate description of molecular-level studies of tissues where the salt solution is an important component. Likewise, the application of this model could be done to study the frequency-dependent and the concentration-dependent behaviors for the microwave conductivity of the other electrolyte solutions in a wide range of concentrations and to estimate their several important parameters. However, the dependence of the microwave conductivity on the temperature of the sodium chloride solution has not been examined by the Drude-jellium model. This matter is rather interesting for our future research.

References

[1] Ding N, Wang X L and Wang J 2006 *Desalination* **192**(1-3)18
[2] Pucihar G, Kotnik T, Miklavcic D and Teissie J 2008 *Biophysical Journal* **95**(6) 2837
[3] Liu D S, Astumiang R D and Tsong T Y 1990 *J. Biol. Chem.* **265** (13) 7260
[4] Gavish N and Promislow K 2016 *Phys. Rev. E* **94**(1-1) 012611
[5] Kropman M F and Bakker H J 2001 *Science* **291**(5511) 2118
[6] Qiao W, Yang K, Thoma A and Dekorsy T 2012 *Journal of Infrared, Millimeter, and Terahertz Waves* **33**(10) 1029
[7] Kuang W and Nelson S O 1988 *ASAE* **41**(1) 173
[8] Dean D A, Ramanathan T, Machado D and Sundararajan R 2008 *J. Electrostat* **66**(3-4) 165
[9] Challis L J 2005 *Bioelectromagnetics* **7** 98
[10] Li S, Li S, Anwar S, Tian F, Lu W and Hou B 2014 *Session 2A0 - PIERS Proceedings* 657
[11] Bulavin L A, Alekseev A N, Zabashta Yu F and Tkachev S Yu 2011 *Ukr. J. Phys.* **56**(6) 547
[12] Pines D 1963 *Elementary Excitations in Solids* (Benjamin, New York)
[13] Apostol M and Preoteasa E 2008 *Phys. Chem. Liquids* **46**(6) 653
[14] Fox M 2010 *Optical properties of solids* (2nd Ed. Oxford University Press, Oxford)
[15] Debye P and Hückel E 1923 *Phys. Z.* **24** 185
[16] Ramos R A 2013 *International Journal of Engineering and Applied Sciences* **2**(3) 29
[17] McKendricka A G and Kesava Pai M 1911 *Proc. R. Soc. Edinb.* **31** 649
[18] Walker S H and Duncan D B 1967 *Biometrika* **54** 167
[19] Di Crescenzo A and Martinucci B 2010 *J. Appl. Prob.* **47** 84
[20] Ayres R U 1989 *Technological Transformations and Long Waves* (International Institute for Applied Systems Analysis:Laxenburg, Austria)
[21] Bod R, Hay J and Jennedy S (eds) 2003 *Probabilistic Linguistics* (Cambridge, MA: MIT Press)

IOP Conf. Series: Journal of Physics: Conf. Series **1034** (2018) 012007 doi:10.1088/1742-6596/1034/1/012007

Polar charges effect on multisubband electron mobility in the semiparabolic quantum wells based on AlN/AlGaN/AlN

Pham Thi Bich Thao[1,2], Nguyen Nhut Tuan Hung[1], Tieu Tin Nguyen[1], Nguyen Thanh Tien[1]

[1]College of Natural Science, Can Tho University, 3-2 Road, Can Tho City, Vietnam.
[2] Graduate University of Science and Technology, Vietnam Academy of Science and Technology, 18 Hoang Quoc Viet, Cau Giay, Ha Noi, Vietnam.

E-mail: ptbthao@ctu.edu.vn

Abstract. This work presents multisubband electron mobility in the polarization semi-parabolic quantum wells (SPQWs) AlN/$Al_xGa_{1-x}N$/AlN. First, a theory of multisubband mobility between the lowest two subbands under a uniform external electric field is studied. By using the variation method, the one-dimensional Poisson and Schrödinger equations have been solved within a finite potential barrier model and a bent band figured by all confinement sources (realistic model). Then, we computed and discussed about the effective confining potential profile, the wave function and multisubband mobility. Our result shows that the positive interface polarization charges effect on the distribution of the two-dimensional electron gas (2DEG) in SPQWs so it has great influences on multisubband mobility.

1. Introduction

In recent years, asymmetric quantum well structures have been continued to be a great deal model for study properties of low dimensional structures [1-4]. These structures also exhibit interesting applications in electronic and optoelectronic devices, such as high electron mobility transistor (HEMT), semiconductor lasers, optical switching, infrared photodetectors, and so on [5-9]. In recent work, effect of asymmetry on the electronic properties such as electron transport and spin transport in quantum wells has also been studied by applying an external electric field perpendicular to the interface plane of the structure [10-12]. Attempts have been made to study the optical properties of asymmetric quantum wells [14].

Literature survey shows that the research in this field is often based on solving Schrödinger equation to find the wave functions and analyzing numerical results. Electron mobility are obtained and the effects of the changing quantum well parameters are investigated. In this way one must restrict himself to well-known infinite potential wells such as parabolic and semi-parabolic QWs [13-15]. It should be noted that the infinite potential wells are not physically feasible. However progress in computational physics with high accuracy permits us to investigate the finite potential wells [16], which are important theoretically and experimentally. Few works have also been reported on electron transport in GaAs/AlGaAs and strained layer asymmetric quantum well structures [17, 18]. In these structures the electron mobility is studied by considering lowest subband occupancy. However, in a system, where more than one subband is occupied, the intersubband effects play a major role on the subband mobility [13-16]. Despite the fact that there were some important studied results about the multisubband electron mobility in GaN quantum wells but a systematic study on AlN/$Al_xGa_{1-x}N$/AlN heterostructures (HSs) is still lacking. In this work, we investigate the effects of the widths of the QWs, the aluminum concentration in $Al_xGa_{1-x}N$ and the strength of the applied electric field on the multisubband electron mobility of the polarization SPQWs based on AlN/$Al_xGa_{1-x}N$/AlN HSs. The confinement potential form of the studied system is shown schematically in Figure 1. Especially, we have known that polarization is an important property of a nitride and oxide-based heterostructures [19, 20]. These HSs possess a very high (a real) density of polarization charges bound on the interface ($\sigma \sim 10^{13}$ cm^{-2}). For formation of 2DEG in a polar HS, interface polarization charges take a double role: they are a source to supply carriers (electrons) into the sample, but they also are a source to confine the carriers along the growth direction. So, the polarization confinement effect has to be considered carefully.

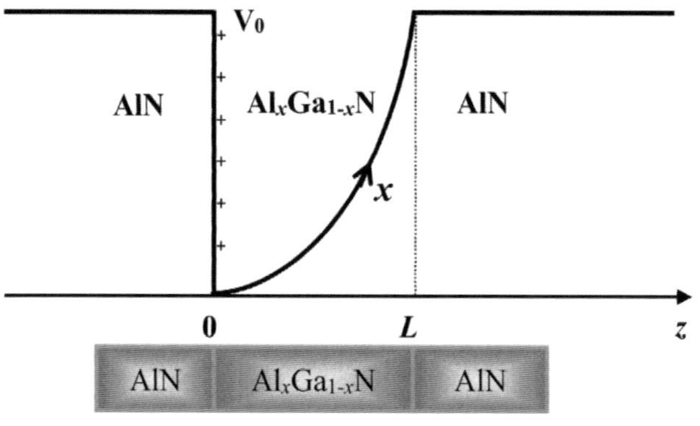

Figure 1. Schematic diagram for a semi-parabolic quantum well.

To do end, the variational method [21, 22] is employed to solve the one-dimensional Poisson and Schrödinger equations for the SPQWs. Then the results are used to calculate the multisubband electron mobility by combining the results of Poisson and Schrödinger to scatterings due to interface roughness and alloy disorder. This paper is organized as follows. In Section 2, we present the theoretical results about how to find the wave functions and eigenvalues of electron in QWs, write down the formalism of the multisubband electron mobility of the system. Section 3 is dedicated to the numerical results for the confining potential forms and the distribution of electrons in the SPQWs. Also, we have investigated the changes of the multisubband electron mobility for the SPQWs. Finally, the conclusions are given in Section 4.

2. Theoretical calculation

2.1. Confining potentials

The two-dimentional electron gas (2DEG) in the semi-parabolic wells is described by the variational wave function [15]:

$$\psi_i(z) = A_i e^{\frac{-\alpha_i^2 z^2}{2}} H_{2i+1}\left[(2\alpha_i + 1)\right] \tag{1}$$

A_i is the normalization contanst, α_i is the typical parameter of band bending. $H_{2i+1}(2\alpha_i + 1)$ denote the Hermite polynominals. The quantum confinement along the z direction is determined by the following Hamiltonian:

$$H = T + V_{\text{tot}}(z) \tag{2}$$

IOP Conf. Series: Journal of Physics: Conf. Series **1034** (2018) 012007 doi:10.1088/1742-6596/1034/1/012007

where T is the kinetic energy, and V_{tot} (z) is the overall confining potential.

Carrier confinement in a polar modulation-doped HS is determined by all confining sources located along the growth direction (z-axis): potential barrier, interface polarization charges, Hartree potential and electric field potential induced by ionized impurities and 2DEG:

$$V_{\text{tot}}(z) = V_b(z) + V_\sigma(z) + V_H(z) + V_{ex}(z) + V_F(z) \quad (3)$$

First, for the potential barrier with a finite height V_0, it can be written:

$$V_b(z) = \begin{cases} \frac{V_0}{L^2} z^2 & 0 \le z \le L \\ V_0 & z < 0; z > L \end{cases} \quad (4)$$

It is well known [20, 23] that due to piezoelectric and spontaneous polarizations in a nitride-based strained HS there exist positive polarization charges bound on the interface. These charges create a uniform normal electric field with the potential given by:

$$V_\sigma(z) = \frac{2\pi e}{\varepsilon}\sigma|z| \quad (5)$$

with σ as the density of polarization charges. ε is the average dielectric constant of the sample.

Next, the Hartree potential induced by the ionized donors and 2DEG in the HS. We consider the non-doping system, so $V_H = V_s$, $V_s(z)$ is confined potential electrons themselves. This is determined according to Poisson's equation:

$$\frac{d^2V_{si}(z)}{dz^2} = -\frac{4\pi e^2}{\varepsilon_a}N_i(z) \quad (6)$$

Here, $N_s(z)$ is the bulk density of electron. It holds

$$N_i(z) = n_i(z)|\psi(z)|^2 \quad (7)$$

n_i is the number of electrons per unit area in the ith subband [19]:

$$n_i = \frac{m^* k_B T}{\pi^- \hbar} \ln\left[1 + \exp\left(E_{Fi} - E_i\right)/k_B T\right] \quad (8)$$

with m^* as the in-plane effective mass of the electron and Fermi level E_{Fi} is given by:

$$E_{Fi} = k_B T \ln\left[\exp\left(\frac{\pi^- \hbar n_i}{m^* k_B T}\right) - 1\right] \quad (9)$$

The potential by the electric field along the growth direction is given by:

$$V_F(z) = |e|\,Fz, \quad (10)$$

where, F is the strength of the applied electric field.

2.2. Total energy per electron in the two lowest subband

We now turn to the total energy per electron when the 2DEG occupying the lowest two subbands. The expectation value of the Hamiltonian given by Eqs. (2) and (3) reads [21]:

$$E_i = \langle T\rangle_i + \langle V_b\rangle_i + \langle V_\sigma\rangle_i + \langle V_s\rangle_i/2 + \langle V_{ex}\rangle_i + \langle V_F\rangle_i, \quad (11)$$

where V_s is the value of V_H for undoped systems, V_{ex} is the exchange correlation interaction and has very small value so it is ignorable. Upon employing the above-derived analytic expressions for the individual confining potentials, we may easily calculate their expectation values with the lowest two subband wave function from Eq. (1). The average energies figuring in Eq. (11) are supplied below.

For the first state,

$$\langle T\rangle_1 = \frac{5\sqrt{\pi}\alpha_1 A_1^2 \hbar^2}{2m_z}, \tag{12}$$

$$\begin{aligned}\langle V_b\rangle_1 &= \frac{V_0 A_1^2}{\alpha_1^2}\left(-4\alpha_1 + 3\sqrt{\pi}\alpha_1\right)\\ &+\frac{V_0 A_1^2}{2L^2\alpha_1^3}e^{-L^2\alpha_1^2}\left[-2\left(4+L\alpha_1\left(5+2L\alpha_1\left(2+L\alpha_1\right)\right)\right)+e^{L^2\alpha_1^2}(8+5\sqrt{\pi}\mathrm{Erf}\left[L\alpha_1\right])\right]\\ &+\frac{V_0 A_1^2}{\alpha_1^2}\left[3\sqrt{\pi}\alpha_1 + 2e^{-L^2\alpha_1^2}\alpha_1\left(2+L\alpha_1\right) - 3\sqrt{\pi}\alpha_1 Erf\left[L\alpha_1\right]\right],\end{aligned} \tag{13}$$

$$\langle V_\sigma\rangle_1 = \frac{16\pi e A_1^2\sigma}{\varepsilon_a \alpha_1^2}, \tag{14}$$

$$\langle V_S\rangle_1 = -\frac{2\pi e^2 A_1^4 n_1}{\varepsilon_a\alpha_1^3}\left(24\pi + 31\sqrt{2\pi}\right), \tag{15}$$

$$\langle V_F\rangle_1 = \frac{4\sqrt{\pi}A_1^2\left|e\right|F}{\alpha_1^2}, \tag{16}$$

and for the second state,

$$\langle T\rangle_2 = \frac{436\sqrt{\pi}\alpha_2 A_2^2 \hbar^2}{m_z}, \tag{17}$$

$$\begin{aligned}\langle V_b\rangle_2 &= \frac{V_0 A_2^2}{\alpha_2}\left(344\sqrt{\pi} - 592\right)\\ &+\frac{V_0 A_2^2}{L^2\alpha_2^3}\left[e^{-L^2\alpha_2^2}\left(916\sqrt{\pi}e^{L^2\alpha_2^2}\mathrm{Erf}\left[L\alpha_2\right] - \left(32L^7\alpha_2^7 + 192L^6\alpha_2^6 + 496L^5\alpha_2^5\right.\right.\right.\\ &\left.\left.\left.+ 832L^4\alpha_2^4 + 1216L^3\alpha_2^3 + 1616L^2\alpha_2^2 + 1832L\alpha_2 + 1616\right)\right) + 1616\right]\\ &+\frac{8V_0 A_2^2 e^{-L^2\alpha_2^2}}{\alpha_2^2}\left[\left(43\sqrt{\pi}\alpha_2 e^{L^2\alpha_2^2} + 2\alpha_2\left(2L^5\alpha_2^5 + 12L^4\alpha_2^4 + 29L^3\alpha_2^3\right.\right.\right.\\ &\left.\left.\left.+ 40L^2\alpha_2^2 + 42L\alpha_2 + 37\right)\right) - 43\sqrt{\pi}\alpha_2 e^{L^2\alpha_2^2}\mathrm{Erf}\left[L\alpha_2\right]\right],\end{aligned} \tag{18}$$

$$\langle V_\sigma\rangle_2 = \frac{3776\pi e A_2^2\sigma}{\varepsilon_a\alpha_2^2}, \tag{19}$$

$$\langle V_S\rangle_2 = -\frac{2\pi e^2 A_2^4 n_2}{\varepsilon_a\alpha_2^3}\left(726528\pi + 358351\sqrt{2\pi}\right), \tag{20}$$

$$\langle V_F\rangle_2 = \frac{1056\sqrt{\pi}A_2^2\left|e\right|F}{\alpha_2^2}. \tag{21}$$

2.3. The electron mobility

The electron mobility μ relation with the transport life time τ is given by:

$$\mu = \frac{e\tau}{m^*} \tag{22}$$

For multisubband with Rth (AD-, SR-, PSR-...) scattering event, the electron mobility μ^P can be written as

$$\mu^P = \sum_i \frac{n_i \mu_i^R}{n_i} \tag{23}$$

The total mobility μ is calculated using Matthiessens relation [24].

When $i = 1$, we can write τ_1 as

$$\frac{1}{\tau_1} = B_{11} \tag{24}$$

where B_{ii} is the intrasubband scattering rate matrix element. However, when we consider two subband ($n = 1, 2$), τ_1 and τ_2 must have intersubband scattering rate matrix elements C_{ij} and D_{ij} [25, 26]:

$$\begin{aligned} \frac{1}{\tau_1} &= \frac{(B_{11} + C_{12})(B_{22} + C_{21}) - D_{12}D_{21}}{(B_{22} + C_{21}) + (E_{F2}/E_{F1})^{1/2} D_{12}} \\ \frac{1}{\tau_2} &= \frac{(B_{11} + C_{12})(B_{22} + C_{21}) - D_{12}D_{21}}{(B_{11} + C_{12}) + (E_{F1}/E_{F2})^{1/2} D_{21}} \end{aligned} \tag{25}$$

here, B_{ii}, C_{ij} and D_{ij} are determined by [25, 26]:

$$\begin{aligned} B_{ii} &= m^* \Big/ \left(\pi\hbar^3\right) \int_0^\pi d\theta \,(1 - \cos\theta) \left|V_{ii}^R\,(q_{ii})\right|^2 \\ C_{ij} &= m^* \Big/ \left(\pi^{-}\hbar\right) \int_0^\pi d\theta \left|V_{ii}^R\,(q_{ij})\right|^2 \\ D_{ij} &= m^* \Big/ \left(\pi^{-}\hbar\right) \int_0^\pi d\theta \cos\theta \left|V_{ii}^R\,(q_{ij})\right|^2 \end{aligned} \tag{26}$$

In this work, interface roughness scattering and alloy disorder scattering were considered.

Interface roughness scattering (PSR)

PSR scattering due to the potential induced by both normal surface roughness, or, strictly speaking, barrier roughness (position fluctuations of the potential barrier), V_{SR}, and polarization roughness (position fluctuations of the polarization charges), V_{PR}:

$$V_{PSR} = V_{SR} + V_{PR} \tag{27}$$

Scattering at rough interfaces can be severe if the 2DEG density is high, since the 2DEG tends to shift closer to the interface as the density increases. The roughness at heterojunction interfaces has been traditionally modelled by a Gaussian autocovariance function. However, in our research, surface roughness potential is determined as following [27]:

$$\left|V_{ij}^{SR}\left(q_{ij}\right)\right|^2 = \pi\Delta^2\Lambda^2 e^{\frac{-q_{ij}^2\Lambda^2}{4}} \left| V_0^2 \sum_{i'j'} \psi_{i'}(z)\psi_{j'}(z)\,\varepsilon_{ij,i'j'}^{-1}\left(q_{ij}\right)\right|^2 \tag{28}$$

Polarization is an important property of nitride- and oxide-based HSs. These HSs possess a very high density of polarization charges bound. It have been recently shown that these charges have a three-fold role similar to the ionized impurities. The polarization charges on a rough interface are a carrier supply source, a confining source, and a scattering mechanism for HSs. Therefore, it has become a key scattering and this project, we can consider polarization scattering potential as following [28-30]:

$$\left|V_{ij}^{PR}\left(q_{ij}\right)\right|^2 = \pi\Delta^2\Lambda^2 e^{\frac{-q_{ij}^2\Lambda^2}{4}} \left| \int_{-\infty}^{+\infty} \pm\frac{2\pi e\sigma}{\varepsilon_a r_\pm} e^{\mp q_{ij} z} \psi_{i'}(z)\psi_{j'}(z)\,\varepsilon_{ij,i'j'}^{-1}\left(q_{ij}\right) dz \right|^2 \tag{29}$$

Here, the subindex + refers again to the potential in the region with $z > 0$ (and $r_+ = r_\varepsilon$), while the subindex − to the one with $z < 0$ (and $r_- = 1$). The upper and lower signs refer to the regions with $z > 0$ and $z < 0$, respectively.

In Eqs (28) and (29), Δ and Λ are square roughness height and a correlation length.

Alloy disorder scattering (AD)

Alloy disorder scattering originates from the randomly varying alloy potential in the barrier. This form of scattering is known [31] to be the mobility-limiting mechanism for 2DEGs confined in an alloy channel such as in heterostructures. In 2DEGs confined in binary wells, alloy scattering occurs as a result of the finite penetration of the 2DEG wavefunction into the barrier. In AlGaN/GaN heterostructures, the large electron effective mass, the high 2DEG density and the large alloy scattering potential all combine to make this form of scattering quite strong in spite of the confinement in the binary semiconductor. With the multisubband, AD scattering can be written [32]:

$$\begin{aligned} \left|V_{ij}^{AD}\left(q_{ij}\right)\right|^2 &= \int_0^L dz \sum_j \left[\frac{\{a\left[x\left(z\right)\right]\}^3 \{\delta V\left(x\right)\}^2 x\left(z\right)\{1-x\left(z\right)\}}{4}\right] \\ &\times \left|\sum_{i'j'} \psi_{i'}\left(z\right)\psi_{j'}\left(z\right)\varepsilon_{ij,i'j'}^{-1}\left(q_{ij}\right)\right|^2, \end{aligned} \tag{30}$$

where $a\left[x\left(z\right)\right]$ and $\delta V \approx \Delta E_{\mathrm{c}}$ are the lattice constant and alloy scattering potential respectively. Two terms are determined [19]:

$$a\left[x\left(z\right)\right] = 3.112\left[x\left(z\right)\right] + 3.189\left\{1-\left[x\left(z\right)\right]\right\}, \quad x\left(z\right) = \frac{z^2}{L^2}$$

$$\Delta E_{\mathrm{c}}\left(x\right) = 0.63\left\{E_{\mathrm{g}}\left(x\right) - E_{\mathrm{g}}\left(0\right)\right\},$$

in which

$$E_g\left(x\right) = 6.28x + 3.42\left(1-x\right) - x\left(1-x\right) \tag{31}$$

Screening effect

The dielectric function $\varepsilon(q_{ij})$ in Eqs. (28), (29) and (30) takes account of the screening of scattering potentials by the 2D carriers. As usual, this is evaluated within the random phase approximation [22]:

$$\varepsilon(q_{ij}) = 1 + \frac{q_s}{q_{ij}} F_{Sij,i'j'}\left(q_{ij}\right)\left[1 - G(q_{ij})\right] \tag{32}$$

Here, $q_s = \frac{2m^* e^2}{\varepsilon_0 \hbar}$ is the inverse 2D Thomas-Fermi screening length, with ε_0 as the dielectric constant of the well layer. The local field corrections due to a many-body exchange effect in the in-plane are quantified by $G(q_{ij}) = \frac{q_{ij}}{2}\sqrt{q_{ij}^2 + k_{Fi}k_{Fj}}$.

q_{ij} is the 2D momentum transfer due to a scattering event in the in-plane (in polar coordinates); $q_{ij} = \sqrt{k_{Fi}^2 + k_{Fj}^2 - 2k_{Fi}k_{Fj}\cos\theta}$; with θ as a scattering angle, with k_{Fi} as the Fermi wave number $k_{Fi} = \sqrt{\frac{2m^* E_{Fi}}{\hbar}}$.

The screening form factor $F_{Sij,i'j'}\left(q_{ij}\right)$ takes account of the extension of particle states along the growth direction.

$$F_{Sij,i'j'}\left(q_{ij}\right) = \int_{-\infty}^{+\infty} dz \int_{-\infty}^{+\infty} dz' |\psi(z)|^2 e^{-q(z-z')} |\psi(z')|^2 \tag{33}$$

3. Numerical Results

We calculated the multisubband electron mobility between the first and second subband in the semiparabolic quantum wells based on $AlN/Al_xGa_{1-x}N/AlN$. More specifically, we changed the parameters of the material system to investigate the multisubband electron mobility, e.g. dependences of the multisubband electron mobility on alloy content x and the applied electric field F.

The physical parameters used in our calculations are chosen as follows [19, 20]: $m_z(Al_xGa_{1-x}N)$ vs x, $m^*(Al_xGa_{1-x}N)$ vs x.

To estimate the role of the confining potentials effect on the electron confinement in quantum wells, we show the partial confining potentials and the total confining potential for QW width $L = 50$Å, applied electric field $F = 100$ kV/cm, and the content of Al in $Al_xGa_{1-x}N$ alloy $x = 0.3$ in Figure 2a.

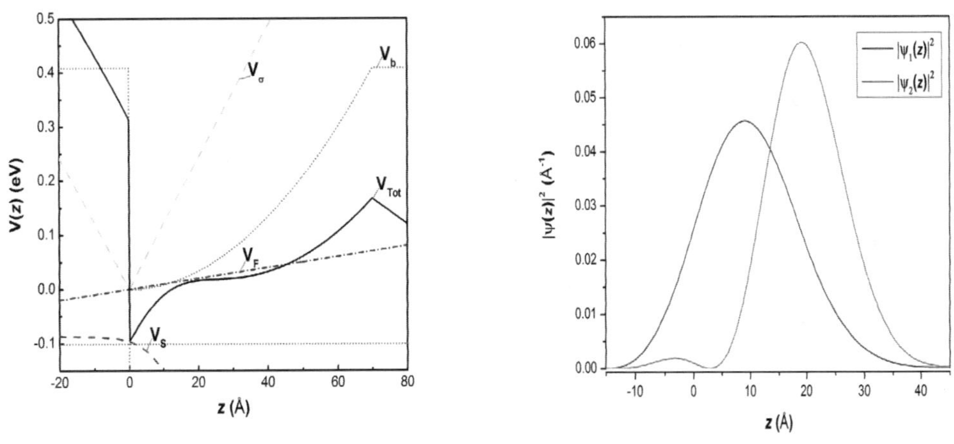

(a) (b)

Figure 2. (a) The partial and total confining potentials. (b) Modul square of the wave fucntion in the lowest two subband.

It can be clearly seen that the polarization induced potential effects the most strong in all of confining potentials. The variation of the semiparabolic confining potential profile by the polarization charges at interface is significantly more the external electric field. The form of the total potential change into the triangle-like at the bottom of quantum well.

In Figure 2b shows modul square of the wave fucntion in the lowest two subband vs the content of Al in $Al_xGa_{1-x}N$ alloy $x = 0.3$, $T = 70$ K, well width $L = 70$ Å without applied electric field. We can see that the peak of two wave function is different, so it is difficult for intrasubband transition to occur. It means intersubband transition still prevail.

IOP Conf. Series: Journal of Physics: Conf. Series **1034** (2018) 012007 doi:10.1088/1742-6596/1034/1/012007

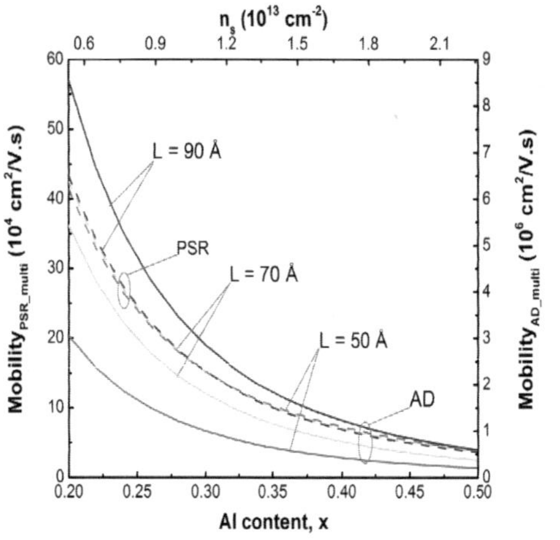

(a) The AD multisubband electron mobility and the PSR multisubband electron mobility.

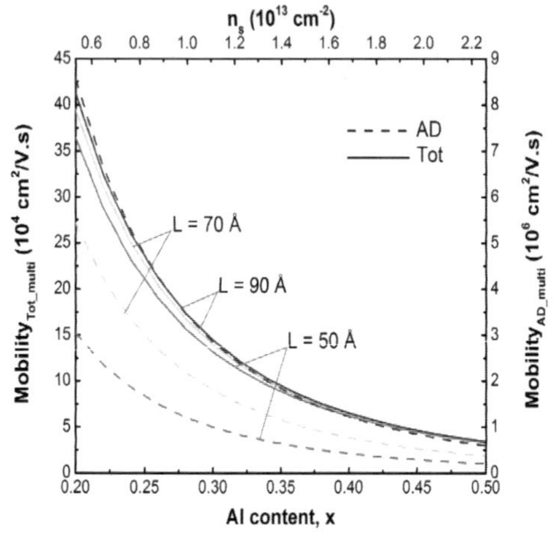

(b) The AD multisubband electron mobility and the total multisubband electron mobility.

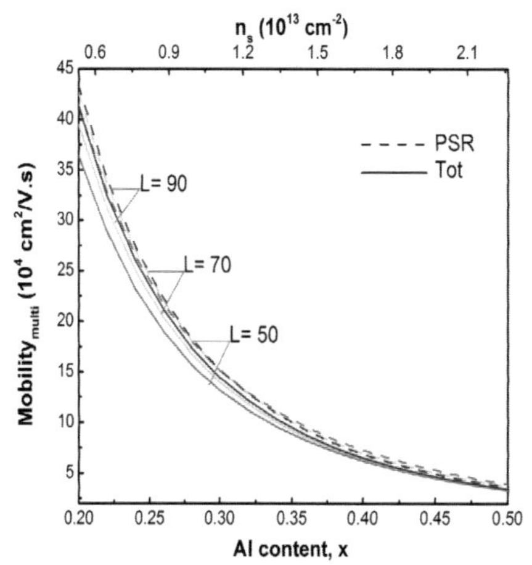

(c) The PSR multisubband electron mobility and the total multisubband electron mobility.

Figure 3. The multisubband electron mobility vs the content of Al.

In Figure 3 shows the multisubband electron mobility vs the content of Al in $Al_xGa_{1-x}N$ alloy $x = 0.2$-0.5, $T = 70$ K, $\Delta = 3$ Å, $\Lambda = 50$ Å, various quantum well widths $L = 50$, 70, and 90 Å without applied electric field.

The multisubband electron mobility vs applied electric field ($F = 0$ kV/cm to 150 kV/cm) with the content of Al in $Al_xGa_{1-x}N$ alloy $x = 0.2$, 0.3, 0.4; $T = 70$ K; $\Delta = 3$ Å ; $\Lambda = 50$ Å; well widths $L = 70$ Å is ploted in Figure 4.

From obtained results in Figure 2, 3 and 4, we may draw the following conclusions.

i) Both alloy disorder scattering and interface roughness scattering effect on the multisubband electron mobility but interface roughness scattering is a key scattering. It means that the multisubband electron mobility is changed mainly by the content of Al in Al $_xGa_{1-x}N$ alloy. Since the density of polarization charges σ is a function of x following $\sigma(x) = 3.075x^2 + 3.70625x$ $(|e|)/cm^2$. Moreover, changing of the content of Al will give different the height barriers V_0 and

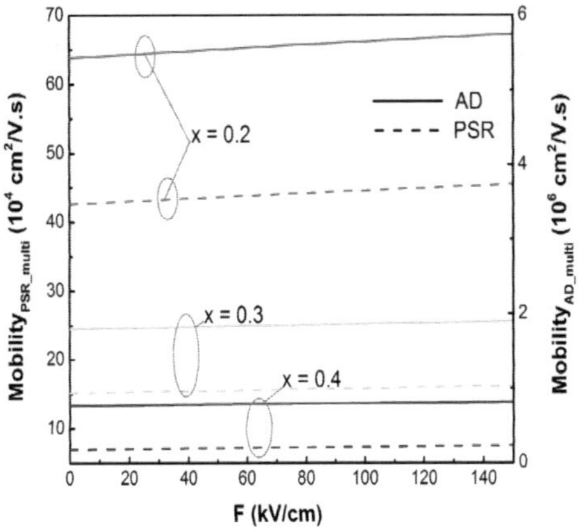

(a) The AD multisubband electron mobility and the PSR multisubband electron mobility.

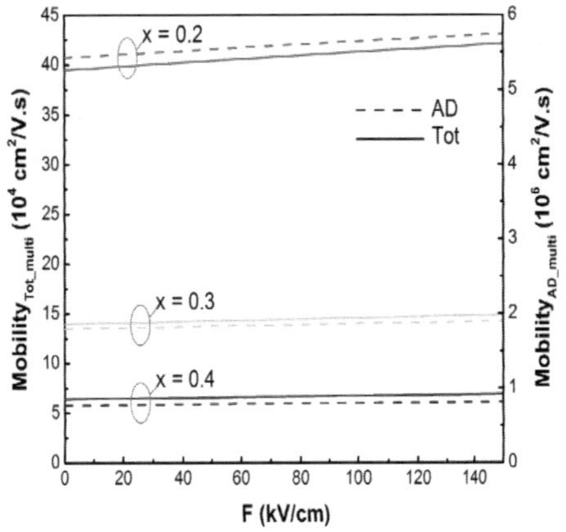

(b) The AD multisubband electron mobility and the total multisubband electron mobility.

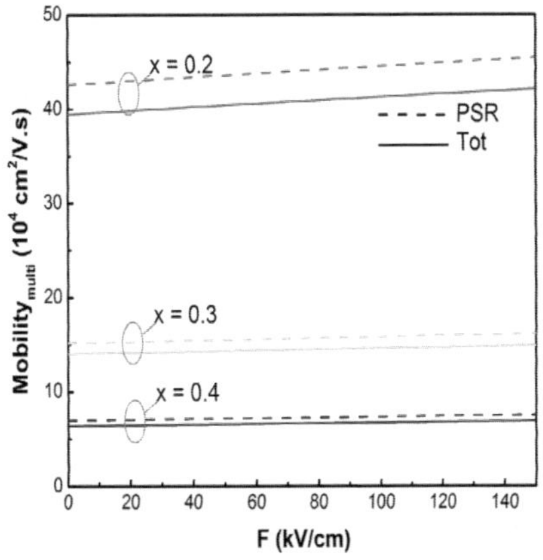

(c) The PSR multisubband electron mobility and the total multisubband electron mobility.

Figure 4. The multisubband electron mobility corresponding to different values F.

the area density of electron.

ii) The density of polarization charges will rise when increasing the content of Al to make 2DEG may be drifted to barrier. As a result, the AD multisubband electron mobility, the PSR multisubband electron mobility and the total multisubband electron mobility decrease due to increasing the content of Al.

iii) The external applied electric field has only slightly effect on the multisubband electron mobility. Therefore, the internal structure parameters (polarization, content of Al in $Al_xGa_{1-x}N$ alloy, the effective width of the potential well) are more important.

4. Conclusions

To summarize, we have determined that both an external applied electric field and the internal structure parameters have an effect on multisubband mobility but the internal structure parameters are more important. This results is a basis for next calculations related to transport and optical properties of the semiparabolic quantum wells based on polar materials. Not long ago, part of the application of the work was used to study linear intersubband optical absorption in the semiparabolic quantum wells based on AlN/AlGaN/AlN under a uniform electric field [33].

Acknowledgments

This research is funded by Vietnam National Foundation for Science and Technology Development (NAFOSTED) under grant number 103.01-2017.74.

References

[1] Khan M A, Pavel A A and Islam N 2013 *IEEE Trans. Nanotechnol.* **12** 521
[2] Sun S L, Fan X L, Feng C F, Wu S Q, Gong G X, Huang and Oh C H 2012 *Opt. Exp.* **20** 8485
[3] Zhou F, Qi Y, Sun H, Chen D, Yang J, Y. Niu Y and Gong S 2013 *Opt. Exp.* **20** (21) 12249
[4] You Bin Yu 2013 *Appl. Mech. Mater.* **389** 1075
[5] Maeda N, Nishida T, Kobayashi N and Tomizawa M 1998 *Appl. Phys. Lett.* **73** 1856
[6] Gaska R, Shur M S, Bykhovski A D, Orlov A O and Snider G L 1999 *Appl.Phys. Lett.* **74** 287
[7] Pearton S J, Zopler J C and Ren F 1999 *J. Appl. Phys.* **86** 1
[8] Karabulut I and Baskoutas S 2008 *J. Appl. Phys.* **103** 073512
[9] Liu C H and Xu B R 2008 *Phys. Lett.* A **372** 888
[10] Lima F M S, Fonseca A L A and Nunes O A C 2002 *J. Appl. Phys.* **92** 5296
[11] Sahu T and Alan Shore K 2010 *J. Appl. Phys.* **107** 113708
[12] Wang G, Liu B L, Balocihi , Renucci P, Zhu C R, Amand T, Fontaine C and Marie X 2013 *Nat. Commun.* **4** 2372
[13] Zhang L 2004 *Opt. Quantum Electron.* **36** 665
[14] Zhang C J and Guo K X 2007 *Physica* B **387** 276
[15] Wang G 2014 *Optik* **125** 2374
[16] Tien N T, Thao D N, Thao P T B, Quang D N 2015 *Physica* B **479** 62
[17] Brunhes T, Boucaud P, Sauvage S 2000 *Phys. Rev.* B **61** 5562
[18] Zhao X F and Liu C H 2006 *Eur. Phys. J.* B **53** 209
[19] Jena D and Wood C 2008 *Polarization Effects in Semiconductors: From Ab Initio Theory to Device Applications* (New York: Springer)
[20] Yao T and Hong S K 2009 *Oxide and Nitride Semiconductors* (Berlin Heidelberg: Springer)
[21] Ando T, Fowler A B and Stern F 1982 *Rev. Mod. Phys.* **54** 437
[22] Quang D N, Tuan L and Tien N T 2008 *Phys. Rev.* B **77** 125326
[23] Ambacher O, Foutz B, Smart J, Shealy J R, Weimann N G, Chu K, Murphy M, Sierakowski A J, Shaff W J, Eastman L F, Dimitrov R, Mitchell A and Stutzmann M 2000 *J. Appl. Phys.* **87** 334
[24] Schubert E F 2006 Ph*ysical Foundations of Solid-State Devices* (New York: Rensselaer Polytechnic Institute, Troy)
[25] Sahu T and Alan Shore K 2009 *Semicond. Sci. Technol.* **24** 095021
[26] Sahu T, Palo S and Panda A K 2013 *J. Appl. Phys.* **113** 083704
[27] Das S, Nayak R K, Panda A K and Sahu T 2014 *Superlat. Micro-struct.* **66** 39
[28] Quang D N, Tung N H and Tien N T 2011 *J. Appl. Phys.* **109** 113711
[29] Tien N T, Thao D N, Thao P T B and Quang D N 2016 *J. Appl. Phys.* **119** 214304
[30] Ando T, Fowler A B and Stern F 1982 *Rev. Mod. Phys.* **54** 437
[31] Bastard G D *Wave-Mechanics applied to Semiconductor Heterostructures* (France: Les Editions de Physique, 1st ed. 27)
[32] Sahoo N and Sahu T 2014 *J. Semicond.* **35**(1)
[33] Tien N T, Hung N N T, Nguyen T T and Thao P T B 2017 *Physica* B **519** 63

IOP Conf. Series: Journal of Physics: Conf. Series **1034** (2018) 012008 doi:10.1088/1742-6596/1034/1/012008

Effect of exact thermal pairing on nuclear level density

Le Thi Quynh Huong[1,2,3], **Nguyen Quang Hung**[1], **Le Tan Phuc**[1,3], **and Nguyen Hoang Tung**[1,3]

[1] Institute of Fundamental and Applied Sciences, Duy Tan University, 3 Quang Trung, Danang city, Vietnam

[2] Department of Natural Science and Technology , University of Khanh Hoa, Nha Trang City, Khanh Hoa Province, Vietnam

[3] Faculty of Physics and Engineering Physics, Ho Chi Minh University of Science, Ho Chi Minh City, Vietnam

E-mail: `lethiquynhhuong@ukh.edu.vn` (L.T.Q.H)

Abstract.
Nuclear level density is studied within a microscopic approach, which is derived based on the exact thermal pairing of the pairing Hamiltonian for the truncated single-particle levels around the Fermi surface in combination with the finite-temperature independent-particle model for the single-particle levels outside the truncated space. The numerical calculations are carried out for ^{170,171,172}Yb isotopes, whose experimental data are available. The results obtained show that the exact thermal pairing is indeed very important for the valid description of nuclear level density in the low and intermediate regions of excitation energy.

1. Introduction

Nuclear level density (NLD), which was first introduced by Hans Bethe in 1936, is defined as the number of excited levels per unit of excitation energy [1]. This quantity has been investigated within various experimental and theoretical studies such as slow neutron resonances at the neutron binding energy and cumulative numbers of discrete levels at low excitation energy [2, 3]. The NLD plays an important role in the understanding of several properties of atomic nuclei such as pairing correlation, nuclear temperature, entropy, heat capacity, etc [4].

Theoretically, the NLD has been extensively calculated within a number of phenomenological and microscopic models. Two effective phenomenological models, which have been popularly employed in the study of NLD, are the back-shifted Fermi gas (BSFG) [5] and constant temperature model (CTM) [6]. These models are both derived based on the Fermi gas picture in which nucleons are supposed to be the non-interacting Fermi particles moving in a square-well potential. To describe the shell and pairing effects, several empirical parameters extracted from the fitting to the experimental data are introduced, for example, the back-shifted energy, shell correction, level density parameter, and pairing energy. Consequently, the quality of NLD predictions depends on the reliability of systematic estimates of those empirically adjusted parameters. When no experimental data is available, the results obtained within those methods are certainly not reliable. In this case, the microscopic models should be employed instead of the phenomenological ones.

Within the microscopic approaches, the effects of superfluid pairing, deformation, vibration, rotation, and shell structure, which are altogether important for the description of NLD, should be either microscopically or empirically included. The NLD can be well reproduced within the finite-temperature shell model Monte Carlo (SMMC) [7]. However, this method is very time-consuming, especially when it is applied to heavy nuclei. Other two microscopic approaches, which have been widely used to estimate the NLD, are the Hartree-Fock Bardeen-Cooper-Schrieffer (HFBCS) [8] and Hartree-Fock-Bogoliubov plus combinatorial method (HFBC) [9]. Although, these two approaches are considered as the most microscopic approaches to NLD up to date, they both violate the conservation of particle number, which is important for the description of light and neutron/proton-rich nuclei [8, 9, 10, 11, 12]. Therefore, in order to have a good description of the experimental data, the NLDs obtained within these approaches have to be renormalized using two phenomenological parameters, whose values are extracted from the fitting to the existed data, namely the NLD data obtained from the analysis of the cumulative numbers of discrete levels at low excitation energy and the data calculated from the average neutron resonance spacing at the neutron binding energy (See e.g., Eq. (9) of Ref. [13]).

The present paper will review our recent microscopic approach to the NLD proposed in Ref. [14]. This approach is derived based on the exact thermal pairing solution to avoid the particle-number conserving problem for the truncated single-particle levels around the Fermi surface in combination with the independent-particle model at finite temperature for the levels outside the truncated space. Moreover, more details on the method, which have not written in Ref. [14] because of its limitation on the number of pages, will be presented in the present paper. In particular, we will present and discuss the results of the thermodynamic quantities such as total energy, heat capacity, free energy and consequently explain the contribution of these quantities to the prediction of the NLD. The comparison of the NLD obtained within our method with that obtained within the independent-particle model at finite temperature (without pairing) and the HFBCS will also be additionally discussed within the present paper.

2. Formalism

2.1. Exact solution of the nuclear pairing at finite temperature (EP)

The paring Hamiltonian describes a system of N particles with single-particle energies ϵ_j interacting via a monopole pairing interaction G as [15]

$$H_p = \sum_{jm} \epsilon_j (a^\dagger_{jm} a_{jm} + \widetilde{a}^\dagger_{jm}\widetilde{a}_{jm}) - G \sum_{jm,j'm'} a^\dagger_{jm}\widetilde{a}^\dagger_{jm}\widetilde{a}_{j'm'} a_{j'm'} \ , \quad (1)$$

with $\widetilde{a}_{jm} \equiv (-1)^{j-m} a_{j-m}$.

In Eq. (1), $a^\dagger_{jm}(a_{jm})$ is the creation (annihilation) operator of a nucleon with angular momentum j, spin projection m_j, and energy ϵ_j, whereas $\widetilde{a}^\dagger_{jm}(\widetilde{a}_{jm})$ is the creation (annihilation) operator of a nucleon with angular momentum j but spin projection -m_j having the opposite direction.

By using the quasi-spin operators of the form

$$L^-_j = \frac{1}{2}\sum_m \widetilde{a}_{jm} a_{jm} \ ,$$

$$L^+_j = (L^-_j)^\dagger = \frac{1}{2}\sum_m a^\dagger_{jm}\widetilde{a}^\dagger_{jm} \ , \quad (2)$$

$$L^z_j = \frac{1}{2}\sum_m (a^\dagger_{jm} a_{jm} - \frac{1}{2}) = \frac{1}{2}(N_j - \Omega_j) \ , \quad (3)$$

IOP Conf. Series: Journal of Physics: Conf. Series **1034** (2018) 012008 doi:10.1088/1742-6596/1034/1/012008

the Hamiltonian (1) can be rewritten as [15]

$$H = \sum_j \epsilon_j \Omega_j + 2\sum_j \epsilon_j L_j^z + G \sum_{jj'} L_j^+ L_{j'}^- , \tag{4}$$

where $N_j = \sum_m a_{jm}^\dagger a_{jm}$ is the particle-number operator and $\Omega_j = j+1/2$ is the level degeneracy. It is noticed that the operators L_j^+, L_j^-, and L_j^z close an SU(2) algebra of angular momentum. Therefore, the pairing problem with the Hamiltonian (1) can be solved exactly by diagonalizing directly a matrix, which is constructed based on the subsets of basis states $|k\rangle \equiv |\{s_j\}, \{N_j\}\rangle$. The latter are expressed in terms of the partial occupation number $N_j \equiv (L_j^z + \Omega_j)/2$ and the partial seniority $s_j \equiv \Omega_j - 2L_j$ (the number of unpaired particles) of the single-particle orbital. Using these basis states, the diagonal and off-diagonal elements of the Hamiltonian matrix are given as [15]

$$\begin{aligned}
&\langle \{s_j\}, \{N_j\} |H| \{s_j\}, \{N_j\}\rangle = \sum_j \left(\epsilon_j N_j + \frac{G}{4} (N_j - s_j)(2\Omega_j - s_j - N_j + 2) \right) , \\
&\langle \{s_j\}, ...N_j + 2, ...N_{j'} - 2, ... |H| \{s_j\}, ...N_j, ...N_{j'}, ...\rangle \\
&= \frac{G}{4} \left[(N_{j'} - s_{j'})(2\Omega_{j'} - s_{j'} - N_{j'} + 2)(2\Omega_j - s_j - N_j)(N_j - s_j + 2) \right]^{1/2} .
\end{aligned} \tag{5}$$

Diagonalizing the matrix (5) for each total seniority $S = \sum_j s_j$, one obtains the exact eigenvalues $\mathcal{E}_S$ and eigenvectors f_j^S. These exact eigenvalues and eigenvectors are then used to construct the canonical partition function at finite temperature T as [16]

$$Z(T) = \sum_S 2^S \exp\left(-\frac{\mathcal{E}_S}{T} \right) . \tag{6}$$

Consequently, all the thermodynamic quantities such as total energy E, heat capacity C, free energy F, paring gap Δ, and single-particle occupation numbers are calculated based on the partition function as

$$F = -T \ln Z(T) , S = \frac{\partial F}{\partial T}, \tag{7}$$

$$E = F + TS , C = \frac{\partial E}{\partial T}, \tag{8}$$

$$\Delta = \sqrt{-GE_{pair}} , E_{pair} = E - 2\sum_j \left[\epsilon_j - \frac{G}{2} f_j \right] f_j, \tag{9}$$

$$f_j = \frac{1}{Z(T)} \sum_S 2^S f_j^S \exp\left(-\frac{\mathcal{E}_S}{T} \right) . \tag{10}$$

2.2. Nuclear level density within the exact paring plus independent-particle model (IPM) at finite temperature (EP+IPM)

Due to the limitation in the size of the matrix to be diagonalized, the EP is carried out only for a limited number of levels around the Fermi surface where pairing is mostly affected. To account for the single-particle levels outside this truncated space, we have employed the independent-particle model at finite temperature (IPM) [17, 18, 19], which treats the nucleons inside the nucleus similar to independently moving particles, that is, without pairing interaction. The occupation numbers of these independent nucleons are approximated well by the Fermi-Dirac distribution as

$$f_j = \frac{1}{1 + \exp\left(\frac{\epsilon_j - \lambda}{T} \right)} , \tag{11}$$

where the chemical potential λ is obtained by solving the equation for the particle number

$$N = 2\sum_j f_j \;, \tag{12}$$

with N being the number of neutrons or protons.

The total EP+IPM partition function is then written as [17, 18, 19]

$$\ln Z'_{EP+IPM} = \ln Z'_{EP} + \ln Z'_{IPM} - \ln Z'_{IPM,tr} \;, \tag{13}$$

where $Z' = Ze^{E_0/T}$ is the excitation partition function with respect to the ground-state energy E_0, Z'_{EP+IPM} is the EP+IPM total partition function, Z'_{EP} is the EP partition function (6) for the truncated single-particle levels, and Z'_{IPM} and $Z'_{IPM,tr}$ are the partition functions calculated within the IPM for the whole and truncated single-particle levels, respectively. Knowing the total partition function, one can easily calculate all the thermodynamic quantities by using Eqs. (7) and (8).

The density of states at a given excitation energy E^* calculated from the inverse Laplace transformation of the partition function is defined as [12]

$$\omega\left(E^*\right) = \frac{e^S}{T\sqrt{2\pi C}} \;, \tag{14}$$

where the excitation energy $E^*(T) = E(T) - E(T=0)$, the total energy $E(T)$, the entropy S and the heat capacity C are obtained from the total partition function (13). The total NLD can be calculated from the state density as

$$\rho\left(E^*\right) = \frac{\omega\left(E^*\right)}{\sqrt{2\pi}\sigma} \;, \tag{15}$$

where σ is the spin cut-off parameter. In axially deformed nuclei, this spin cut-off parameter contains the perpendicular $\sigma_\perp$ and parallel $\sigma_\parallel$ components, whose explicit formulas can be taken from the empirical forms as [5, 20]

$$\sigma_\perp^2 \approx 0.015A^{5/3}T \;, \; \sigma_\parallel = \sigma_\perp\sqrt{\left(3-2\beta_2\right)/\left(3+\beta_2\right)} \;, \tag{16}$$

with A and β_2 being the mass number and the quadrupole deformation parameter.

To take into account the vibrational and rotational degree of freedom, which are not present in the pairing Hamiltonian (1), the NLD (15) is then multiplied with the vibrational k_{vib} and rotational k_{rot} enhancement factors, whose empirical formulas are given as [20, 21, 22]

$$k_{vib} = \exp\left(0.0555A^{2/3}T^{4/3}\right) \;, k_{rot} = \frac{\sigma_\perp^2 - 1}{1 + \exp\frac{E^*-U_C}{D_C}} + 1 \;, \tag{17}$$

with $D_C = 1400\beta_2^2A^{-2/3}$ and $U_C = 120\beta_2^2A^{1/3}$ [20].

The final total NLD obtained within the EP+IPM is [23]

$$\rho\left(E^*\right) = k_{rot}k_{vib}\frac{\omega\left(E^*\right)}{\sqrt{2\pi}\sigma_\parallel} \;. \tag{18}$$

3. Numerical results and discussion

The NLDs are calculated within the EP+IPM for ^{170,171,172}Yb. The results obtained are then compared with the available experimental data given by the Oslo group [24, 25]. The single-particle spectra ϵ_j are taken from the axially deformed Woods-Saxon (WS) potential including the spin-orbit and Coulomb interactions. The parameters of the WS potential are taken from Ref. [26]. The quadrupole deformation parameters β_2 are adjusted to reproduce the experimental ground-state properties including the nuclear binding energy and charge radii. The pairing interaction parameter $G_{N(Z)}$ is as usual adjusted so that the pairing gap $\Delta_{N(Z)}$ obtained within the EP+IPM at $T = 0$ fits the experimental odd-even mass difference. The exact diagonalization of the pairing Hamiltonian is carried out for 12 single-particle levels, in which 6 levels are above and 6 levels are below the Fermi surface. The entire single-particle spectra, which are used for the EP+IPM calculation, contain all the doubly degenerate levels from bottom to a N=126 closed shell.

In Figs. 1(a1)-1(a3), it is clearly seen that the exact neutron (solid lines) and proton (dotted lines) pairing gaps decrease with increasing T and remain finite even at T=3 MeV, well above the critical temperature $T_C \approx 0.57\Delta(T = 0)$, where the pairing gap collapses within the conventional finite-temperature BCS (FTBCS) theory [27, 28, 29, 30, 31]. In ^{171}Yb, a slight increase in the neutron gap (solid lines in Fig. 1(a2)) is seen at low $T < 0.5$ MeV because of the blocking effect causing by the odd neutron [32].

The excitation energies E^* as functions of T obtained within the IPM (dotted lines), EP+IPM (solid lines) in comparison with those obtained within the HFBC (dashed lines), HFBCS (dash dotted lines) are shown in Figs. 1(b1)-1(b3). At low $T < 0.7$ MeV, E^*_{EP+IPM} is always lower than those obtained within other three approaches, whose predictions almost coincide. This difference is due to the fact that the pairing is treated exactly within the EP+IPM while it is only approximated within the HFBC and HFBCS or even is not included as the case of the IPM. In contrast, at $T > 0.7$ MeV, E^*_{EP+IPM} is higher than E^*_{IPM}, E^*_{HFBC}, and E^*_{HFBCS} because in this temperature region, which is above T_C, the pairing gaps obtained within both the HFBC and HFBCS are zero, whereas the EP+IPM gaps still remain finite as shown in Figs. 1(a1)-1(a3). Moreover, it is also seen in that E^*_{HFBC} at $T > 0.7$ MeV is always the lowest one. The reason is that the HFBC is constructed based on the partition function of the incoherent 1 particle - 1 hole (1p1h) state densities built on top of the HFB single-particle spectra. These 1p1h states are certain not enough for the high temperature and/or excitation energy regions. For the latter, all the possible couplings of particle and hole states such as 2p2h, 3p3h, etc., as well as particle-particle (pp) and hole-hole (hh) should be included. Meanwhile, since the EP+IPM is derived based on the exact CE partition function obtained from the direct diagonalization of the matrix elements of the Hamiltonian, which include all the possible couplings between the ph, pp, and ph states, E^*_{EP+IPM} is always higher than E^*_{HFBC} in the high-temperature region.

Shown in the Figs. 1(c1)-1(c3) and 1(d1)-1(d3) are the entropy S and heat capacity C obtained within the IPM and EP+IPM. Due to the presence of the pairing in the entire temperature region, S_{EP+IPM} is always lower than S_{IPM} in which no pairing is included, whereas C_{EP+IPM} is lower (higher) than C_{IPM} at low (high) temperature region. These results indicate that the existence of exact thermal pairing tends to make the system more stable than the cases without pairing.

The results of the thermodynamic quantities shown in Fig. 1 strongly affect the corresponding NLD ρ depicted in Fig. 2. In fact, Fig. 2 shows that without pairing, ρ_{IPM} strongly overestimates the experimental data, whereas with pairing being included, ρ_{EP+IPM}, ρ_{HFBCS}, and ρ_{HFBC} are almost close to the experimental data. Especially, ρ_{EP+IPM} (solid lines) almost coincides with the experimental data in the excitation energy region below the neutron binding energy $E^* < B_n$. ρ_{EP+IPM} also coincides with ρ_{HFBC} for both negative and positive parities. The latter are the NLDs obtained after the normalization to the known data as mentioned

IOP Conf. Series: Journal of Physics: Conf. Series **1034** (2018) 012008 doi:10.1088/1742-6596/1034/1/012008

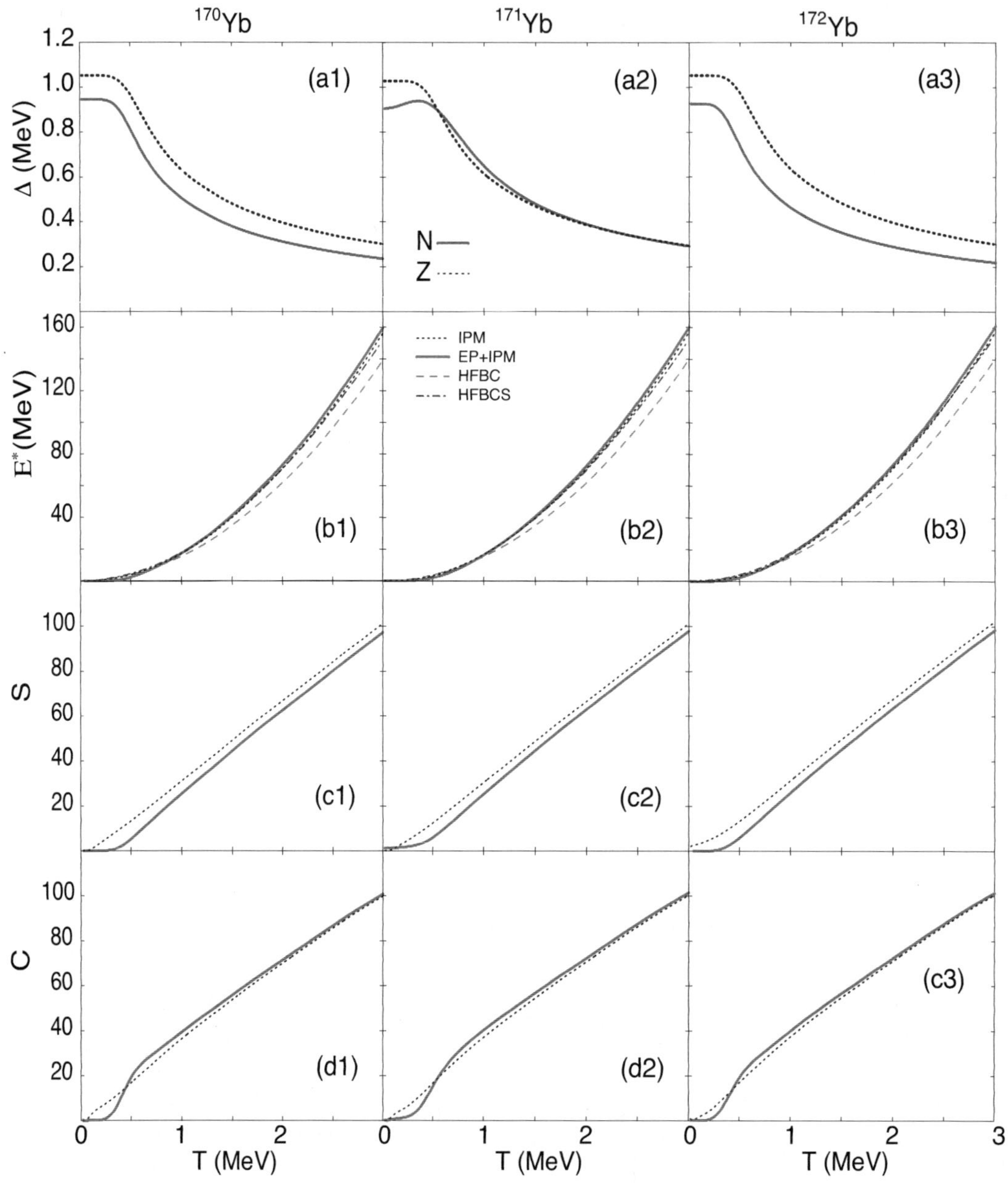

Figure 1. Exact neutron and proton pairing gaps Δ [(a1)-(a3)], excitation energy E^* [(b1)-(b3)], entropy S [(c1)-(c3)], and heat capacity C [(d1)-(d3)] as functions of T obtained within the IPM, EP+IPM as well as the HFBCS and HFBC (negative and positive parities) for ^{170,171,172}Yb. In Figs. [(b1)-(b3)], the predictions within the HFBCS and HFBC are taken from RIPL-2 [33] and RIPL-3 [34] databases, respectively.

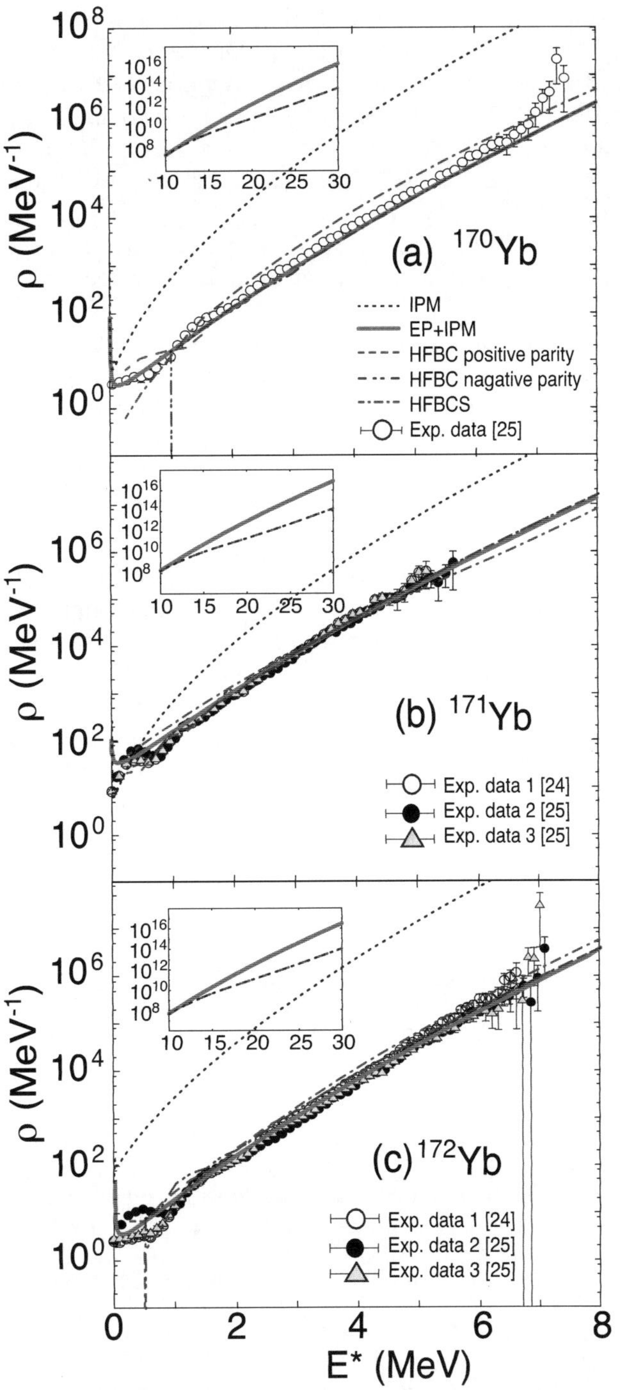

Figure 2. Total level densities ρ as function of E^* obtained within the EP+IPM [14] in comparison with predictions of the IPM, HFBCS [33] and HFBC for the positive and negative parities [34] and the experimental data [24, 25] for ^{170}Yb (a), ^{171}Yb (b), and ^{172}Yb (c).

above. It is also observed in the insets of Fig. 2 that ρ_{HFBC} is much lower than ρ_{EP+IPM} in the high energy region $E^* > B_n$. The reason of this discrepancy is certainly due to the fact that in the high temperature region, the excitation energy obtained within the HFBC, which is constructed based on only the 1p1h state densities, is much lower than that obtained within the EP+IPM as discussed above. These results indicate that the HFBC predictions for the NLD in the high-excitation energy region are still in question. In this case, the EP+IPM predictions seem to be more reliable.

The advantage of the EP+IPM approach proposed here is that it is rather simple, based only on a Woods-Saxon mean-field potential with a set of global parameters for all nuclei. Beyond the Woods-Saxon mean field, the proposed approach uses only two parameters, namely, the monopole pairing interaction parameters G_N (for neutron) and G_Z (for proton), which are fixed at $T = 0$ in order to reproduce the experimental odd-even mass formula at $T = 0$. This means that all of the parameters used in the EP+IPM calculation are fixed at $T = 0$ and no fitting parameters at finite temperature and/or excitation energy are employed, ensuring the microscopic nature of this approach.

4. Conclusions

The present paper reviews our recent microscopic approach published in Ref [14], which offers a valid and reliable description of total nuclear level density. The approach is derived based on the exact solutions of thermal pairing problem in combination with the temperature-dependent independent-particle model. The very good agreement between the results obtained within this approach and the experimental NLD data of ^{170,171,172}Yb isotopes has shown that exact pairing is indeed very important for the description of total nuclear level density in both regions of low and high excitation energies. The advantages of this approach are its microscopic nature and the absence of parameter fitting at finite excitation energies as well as the very short computation time. The latter takes only less than 5 min even for a heavy nucleus, and therefore all the numerical calculations can be performed on a PC.

The present work is supported by the National Foundation for Science and Technology Development (NAFOSTED) of Vietnam through Grant No. 103-04.2017.69.

References

[1] Bethe H A 1936 An Attempt to Calculate the Number of Energy Levels of a Heavy Nucleus *Phys. Rev.* **50** 332
[2] Rauscher T and Thielemann F K 2000 Astrophysical reaction rates from statistical model calculations *At. Data Nucl. Data Tables* **75** 1
[3] Rauscher T, Thielemann F K and Kratz K L 1997 Nuclear Level Density and the Determination of Thermonuclear Rates for Astrophysics *Phys. Rev. C* **56** 1613
[4] Ericson T 1960 The statistical model and nuclear level densities *Adv. Phys.* **9** 425
[5] Dilg W, Schantl W, Vonach H and Uhl M 1973 Level density parameters for the back-shifted fermi gas model in the mass range 40 ¡ A ¡ 250 *Nucl. Phys. A* **217** 269
[6] Gilbert A and Cameron A G W 1965 A composite nuclear-level density formula with shell corrections *Can. J. Phys.* **43** 1446
[7] Alhassid Y, Liu S and Nakada H 1999 Particle-number reprojection in the shell model Monte Carlo method: Application to nuclear level densities *Phys. Rev. Lett.* **83** 4265
[8] Demetriou P and Goriely S 2001 Microscopic nuclear level densities for practical applications *Nucl. Phys. A* **695** 95
[9] Hilaire S and Goriely S 2006 Global microscopic nuclear level densities within the HFB plus combinatorial method for practical applications *Nucl. Phys. A* **779** 63
[10] Ring P and Schuck P 1980 *The nuclear many-body problem* (Heidelberg: Springer Verlag)
[11] Brink D M and Broglia R A 2005 *Nuclear superfluidity: pairing in finite systems* (England: Cambridge University Press)
[12] Bohr A and Mottelson B R 1975 *Nuclear Structure, Vol. I: Single-Particle Motion* (New York: Benjamin)

IOP Conf. Series: Journal of Physics: Conf. Series **1034** (2018) 012008 doi:10.1088/1742-6596/1034/1/012008

[13] Goriely S, Hilaire S and Koning A J 2008 Improved microscopic nuclear level densities within the Hartree-Fock-Bogoliubov plus combinatorial method *Phys. Rev. C* **78** 064307
[14] Hung N Q, Dang N D and Huong L T Q 2017 Simultaneous Microscopic Description of Nuclear Level Density and Radiative Strength Function *Phys. Rev. Lett.* **118** 022502
[15] Volya A et al 2001 Exact solution of the nuclear pairing problem *Phys. Lett. B* **509** 37
[16] Hung N Q and Dang N D 2009 Exact and approximate ensemble treatments of thermal pairing in a multilevel model *Phys. Rev. C* **79** 054328
[17] Alhassid Y, Bertsch G F and Fang L 2003 Nuclear level statistics: Extending shell model theory to higher temperatures *Phys. Rev. C* **68** 044322
[18] Hung N Q and Dang N D 2010 Canonical and microcanonical ensemble descriptions of thermal pairing within BCS and quasiparticle random-phase approximation *Phys. Rev. C* **81** 057302
[19] Hung N Q and Dang N D 2010 Thermodynamic properties of hot nuclei within the self-consistent quasiparticle random-phase approximation *Phys. Rev. C* **82** 044316
[20] Junghans A R et al 1998 Projectile-fragment yields as a probe for the collective enhancement in the nuclear level density *Nucl. Phys. A* **629** 635
[21] Ignatyuk A V 1983 *The Statistical Properties of the Excited Atomic Nuclei* (Moscow: Energoatomizdat)
[22] Iljinov A S 1992 Phenomenological statistical analysis of level densities, decay widths and lifetimes of excited nuclei *Nucl. Phys. A* **543** 517
[23] Capote R et al 2009 RIPL Reference Input Parameter Library for Calculation of Nuclear Reactions and Nuclear Data Evaluations *Nucl. Data Sheets* **110** 3107
[24] Schiller A et al 2001 Critical temperature for quenching of pair correlations *Phys. Rev. C* **63** 021306(R)
[25] Agvaanluvsan U et al 2004 Level densities and γ-ray strength functions in ^{170,171,172}Yb *Phys. Rev. C* **70** 054611
[26] Cwiok S et al 1987 Single-particle energies, wave functions, quadrupole moments and g-factors in an axially deformed woods-saxon potential with applications to the two-centre-type nuclear problems *Comput. Phys. Commun.* **46** 379
[27] Frauendorf S et al 2003 Temperature-induced pair correlations in clusters and nuclei *Phys. Rev. B* **68** 024518
[28] Sheikh J A, Palit R and Frauendorf S 2005 Reappearance of the pairing correlations at finite temperature *Phys. Rev. C* **72** 041301(R)
[29] Mang H J, Rasmussen O and Rho M 1966 Comparisons of BCS Nuclear Wave Functions with Exact Solutions. II. Particle-Number-Conserving Method *Phys. Rev.* **141** 941
[30] Dang N D and Hung N Q 2008 Pairing within the self-consistent quasiparticle random-phase approximation at finite temperature *Phys. Rev. C* **77** 064315
[31] Hung N Q and Dang N D 2008 Pairing in hot rotating nuclei *Phys. Rev. C* **78** 064315
[32] Hung N Q, Dang N D and Huong L T Q 2016 Improved treatment of blocking effect at finite temperature *Phys. Rev. C* **94** 024341
[33] Belgya T et al 2006 *Reference Input Parameter Library (RIPL-2)* (online) Available at: https://www-nds.iaea.org/RIPL-2/ (*Accessed* 14 December 2017)
[34] Capote R et al 2009 *Reference Input Parameter Library (RIPL-3)* (online) Available at: https://www-nds.iaea.org/RIPL-3/ (*Accessed* 14 December 2017)

Computational predictions of the new Gallium nitride nanoporous structures

Le Thi Hong Lien[1], Vu Ngoc Tuoc[1], Do Thi Duong[2] and Nguyen Thu Huyen[2]

[1]Institute of Engineering Physics, Hanoi University of Science and Technology, 1 Dai Co Viet Road, Hanoi, Viet Nam

[2]Hong Duc University, 307 Le Lai, ThanhHoa city, Viet Nam

E-mail: lien.lethihong@hust.edu.vn

Abstract. Nanoporous structural prediction is emerging area of research because of their advantages for a wide range of materials science and technology applications in opto-electronics, environment, sensors, shape-selective and bio-catalysis, to name just a few. We propose a computationally and technically feasible approach for predicting Gallium nitride nanoporous structures with hollows at the nano scale. The designed porous structures are studied with computations using the density functional tight binding (DFTB) and conventional density functional theory methods, revealing a variety of promising mechanical and electronic properties, which can potentially find future realistic applications. Their stability is discussed by means of the free energy computed within the lattice-dynamics approach. Our calculations also indicate that all the reported hollow structures are wide band gap semiconductors in the same fashion with their parent's bulk stable phase. The electronic band structures of these nanoporous structures are finally examined in detail.

Keyword: *Gallium nitride, nanoporous; Structure prediction; DFTB*

1. Introduction

Today, due to the advancement of technology, one can growth the material in layers or even in sub-layer with a desired composition. And moreover, it is expected to be able to manipulate the layers of materials in the desired positions. This opens the door to the creation of porous structures with special shapes.

So far, nanoporous materials have quickly become an important research direction [1, 2]. These nanomaterials are promising for sustainable applications, such as catalysts, gas separation, water purification, and batteries. Gallium nitride equipments is the optimal choice for many emerging power semiconductor applications and rapidly replaces existing silicon technology, providing opportunities for explosive growth over the next decade, including switching conversion in smart grid systems, renewable energy systems, RF radio frequencies), power supply modules for motors/cars, and satellite communications systems, radar technology and space electronics. In the field of semiconductor optics, Gallium nitride GaN [3, 4, 5] is increasingly used in LEDs, laser diodes and optical converters because of its stronger emission characteristics than Si, SiC, GaAs and GaP. AlGaN combined with GaN [6], is used in the semiconductor optical is designed for applications that ultraviolet light brightness extreme operating at wavelengths <400 nm, for example, Green, UV and LED white.

In this paper, we propose a plan for the design of nanoporous structures. Our approach depends on the arrangement of the hexagonal hollow size into the block GaN, keeping the chemical composition unchanged.

2. Computational details

2.1. Theoretical structure prediction approach

In this section, we will briefly introduce the theory of predictive structures we perform. The approach here is top to bottom. From a bulk super cell of GaN, we have created hexagonal holes, from the orthogonal system. This structure will be optimized to get the primitive cells. Then, the structure is run to recover to get the final structure (see Figure 1). The nanostructures reported here are characterized by their walls and their hollow dimensions. The hexagonal hollow size, k, is measured in cubic hexagonal size units. Hollow walls are defined as the thickness between the voids, which may be walls (SW), two walls (DW), three walls (3W) and four walls (4W). In each of the structures of the wall we create different porous structures for study and evaluation. In each of them, the structure of the wall we create a different gap size structures to study and evaluate. Thus, we construct each corresponding structure in the form nW-k where n is the number of walls, k is the hexagonal hollow size. For example, a structure called DW-3 would have will have hollows of size 3 which are spaced by two walls of hexagonal blocks. To illustrate, the series is shown in Figures 1, 2, 3, 4, 5.

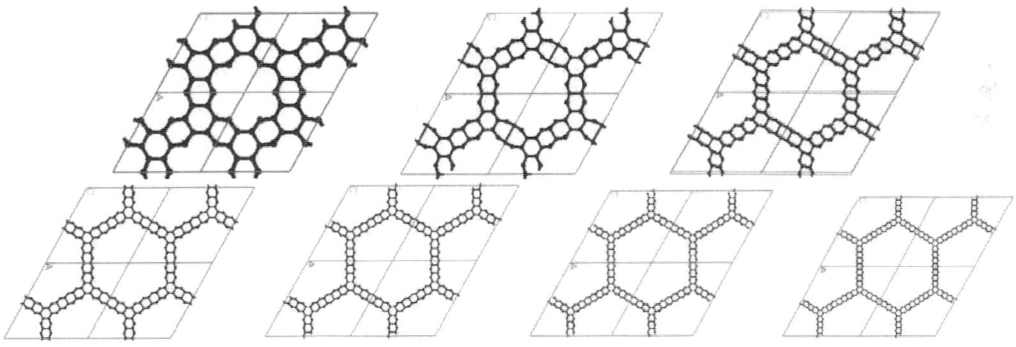

Figure 1. Single-walled structures were investigated respectively as SW-2, Sw-3,...., SW-8.

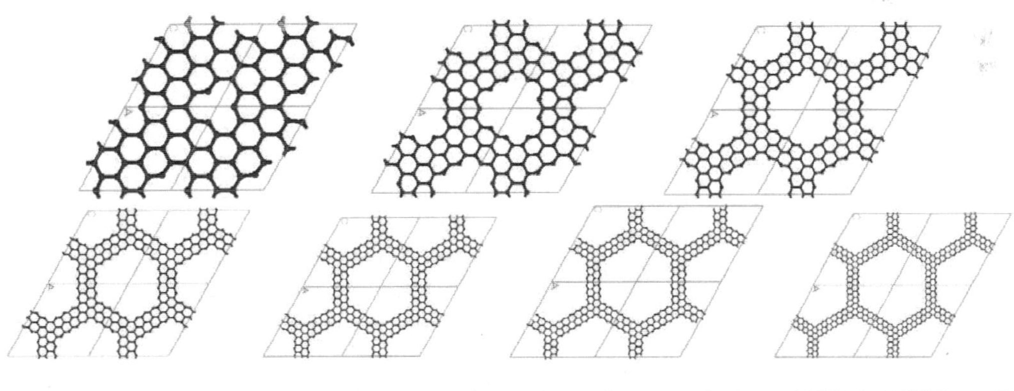

Figure 2. Double - walled structures were investigated respectively as DW-2, DW-3,..., DW-8.

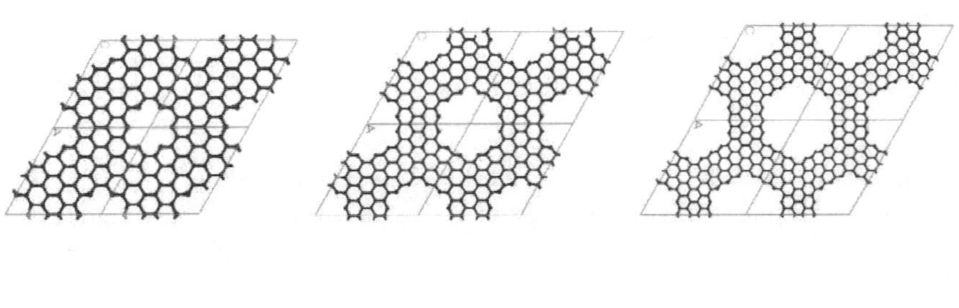

IOP Conf. Series: Journal of Physics: Conf. Series **1034** (2018) 012009 doi:10.1088/1742-6596/1034/1/012009

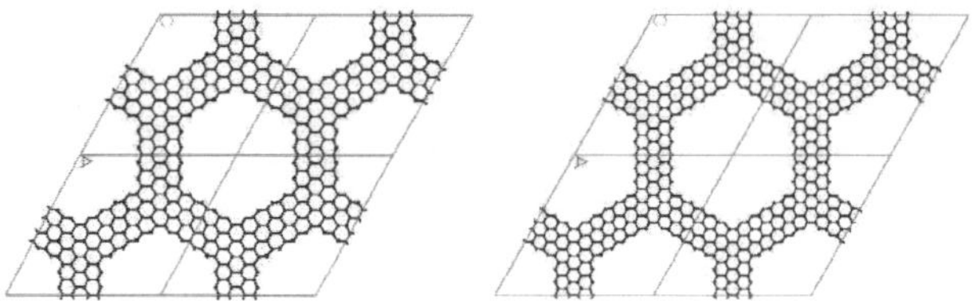

Figure 3. Three - walled structures were investigated respectively as 3W-2, 3W-3,..., 4W-6.

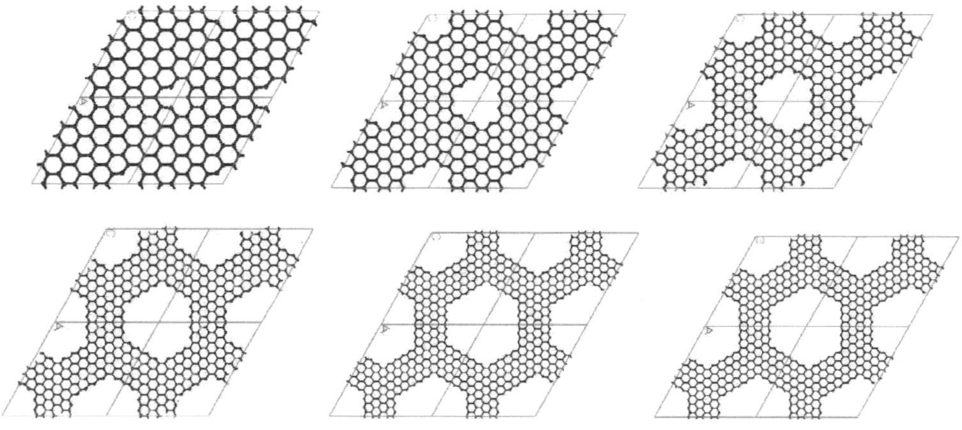

Figure 4. Four- walled structures were investigated respectively as 4W-2, 4W-3,.... 4W-7.

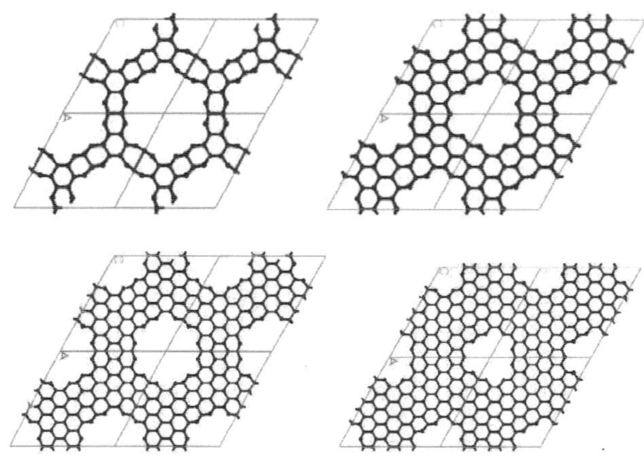

Figure 5. Four structures with cavity radii but the number of walls varies from 1 to 4.

2.2. Density functional based tight binding method

Density Functional based tight binding (DFTB) [7,8] is an approximate approach density functional theory (DFT) that extends the computational scope of DFT tools because they have roughly equivalent levels of accuracy but have computational time faster than the exponent DFT. Thus, it allows study the electronic structure of large systems that cannot be exploited with traditional DFT. However, DFTB requires pre-set parameter files for all pairs of atoms in the compound.

Our calculations are done in a tight-fitting approach (DFTB) [9-12]. This method is based on the second-order extension of the spin-based Kohn-Sham total energy function involving a defined reference charge and the magnetization density. You can find more details on DFTB + in [13,14] and the references in it. In our calculation, the parameter and its conversion properties have been successfully applied in some previous DFTB projects [15, 16]. For exchange-correlation energy we use the functions of Perdew-Burke-Ernzerhof (PBE) [17] and Heyd-Scuseria-Ernzerhof (HSE06) [18].

3. *Results and dicussions*

3.1. Phase stability

For estimation nanoporous phase's stability, we first calculated the total binding energy per Zn-O pair vs. volume for a series of uniformly strains around the relaxed lattice volume. Then the bulk modulus (BM) are obtained by nonlinearly fitting above data series to the third order Birch-Murnaghan equation of states (EOS) as performed in Ref. [19, 20]. The results are presented in Figure 6 as well as listed in the Table of characteristics. It should be noted that, during the optimization process, all the designed nanoporous phases are found to be energetically stable and therefore could survive without structural collapse, showing that these phases are physically relevant and might actually be nano-lattice porous phases for GaN. Also these all are suffered from surface relaxation due to the dangling bonds exposed at the internal hollow surface. The portion dangling bonds [21, 22, 23, 24] can be examined by considering the *average coordination number* (see Table of characteristics), which in some sense reflect the *surface-to-volume* ratio. For all series, this parameter is below four, which is the typical value of wurtzite coordination number and are varying with different hollow and wall sizes as decreasing while the hollow radius increase (see more in Tables).

Structures	**GaN-Hex2-SW**	**GaN-Hex3-SW**	**GaN-Hex4-SW**	**GaN-Hex5-SW**	**GaN-Hex6-SW**	**GaN-Hex7-SW**	**GaN-Hex8-SW**
Mass Density (gcm^{-3})	4.692	3.544	2.787	2.301	1.957	1.288	1.505
Volume/at ($Å^3$/at)	14.817	19.618	24.941	30.207	35.516	53.985	46.182
Part.density ($10^{23}cm^{-3}$)	36.675	60.510	84.401	108.331	132.282	156.185	180.217
Coord. Number	0.00 3.00	0.00 3.00	0.00 3.00	0.00 3.00	0.00 3.00	0.00 1.69	0.00 3.00
Crystall structure	Hex	hex	hex	hex	hex	hex	hex
Symmetry groups	P31M (C3V-2)	P31M (C3V-2)	P31M (C3V-2)	P31M (C3V-2)	P31M (C3V-2)	P31M (C3V-2)	P31M (C3V-2)
Unit cell (Atoms)	36	60	84	108	132	156	180
Lattice param. Å (a-c)	10.917927 5.167124	16.246624 5.196454	21.596548 5.186669	26.944416 5.189022	32.294316 5.190745	37.644865 5.191877	42.994103 5.192761
Aver.bond Å: Ga-Ga;Ga-N	1.866	1.835	1.832	1.827	1.824	1.942	1.820
Aver.Angle Ga-N-Ga/N-Ga-N	112.03 114.19	114.39 115.84	114.25 115.87	114.58 116.16	114.79 116.33	107.79 108.18	115.04 116.53
Band gap eV	3.4014	3.6735	3.8912	3.8640	3.8368	3.8096	3.7824
UC Connolly sur. area $Å^2$	91.98	193.43	299.91	404.31	508.85	613.33	717.80

IOP Conf. Series: Journal of Physics: Conf. Series **1034** (2018) 012009 doi:10.1088/1742-6596/1034/1/012009

UC Pore Volume Å^3	428.90	722.12	1003.16	1288.55	1574.65	1860.20	2147.61

Table 1. Calculated characteristics of all studied phase and most common phase of GaN–SW.

Structures	**GaN-hex-DW221**	**GaN-hex-DW331**	**GaN-hex-DW441**	**GaN-hex-DW551**	**GaN-hex-DW661**	**GaN-hex-DW771**	**GaN-hex-DW881**
Mass Density (gcm^{-3})	6.066	4.786	3.882	3.246	2.783	2.433	2.141
Volume/at (Å^3/at)	11.460	14.525	17.910	21.419	24.980	28.568	32.468
Part.density (10^{23}cm^{-3})	46.873	82.688	118.558	154.467	190.400	226.350	262.308
Coord. Number	0.00 3.87	0.29 3.34	0.00 3.24	0.00 3.18	0.00 3.15	0.00 3.12	0.92 3.14
Crystall structure	Hex	hex	hex	hex	hex	hex	hex
Symmetry groups	P31M (C3V-2)	P31M (C3V-2)	P31M (C3V-2)	P31M (C3V-2)	P31M (C3V-2)	P31M (C3V-2)	P31M (C3V-2)
Unit cell (atoms)	46	82	118	154	190	226	262
Lattice param. Å (a-c)	10.862025 5.159381	16.318867 5.164204	21.73099 5.16752	27.144933 5.169024	32.559 5.169904	37.978984 5.170486	43.384599 5.170661
Aver. Bond Å: Ga-Ga;Ga-N	1.912	2.927 1.894	1.884	1.878	1.874	1.872	2.923 1.876
Average angle Ga-N-Ga/N-Ga-N	109.69 109.84	110.56 111.39	111.16 112.12	11.52 112.55	111.76 112.84	111.93 113.05	111.52 113.48
Band gap eV	2.0953	3.5103	3.4558	3.4558	3.4558	3.4831	3.4831
UC Connolly surf. area Å^2	9.84	143.78	249.09	355.29	461.39	567.57	637.46
UC Pore Volume Å^3	524.19	945.56	1366.59	1784.59	2203.95	2623.78	3045.44

Table 2. Calculated characteristics of all studied phase and most common phase of GaN-DW.

Structures	**GaN-Hex2-3W331**	**GaN-Hex3-3W441**	**GaN-Hex4-3W551**	**GaN-Hex5-3W661**	**GaN-Hex6-3W771**	**GaN-Hex7-3W881**
Mass Density (gcm^{-3})	5.712	4.735	4.052	3.514	3.100	2.771
Volume/at (Å^3/at)	12.171	14.682	17.157	19.784	22.426	25.084
Part.density (10^{23}cm^{-3})	96.822	144.681	192.583	240.505	288.446	336.399
Coord. Number Ga-Ga; Ga-N	0.50 3.87	1.22 3.67	0.00 3.56	0.00 3.50	0.00 3.46	0.00 3.43
Crystall structure	hex	hex	hex	hex	hex	hex

Symmetry groups	P31M (C3V-2)	P31M (C3V-2)	P31M (C3V-2)	P31M (C3V-2)	P31M (C3V-2)	P31M (C3V-2)
Unit cell (atoms)	96	144	192	240	288	336
Lattice param. Å (a-c)	16.205705 5.136579	21.720764 5.162572	27.142431 5.163567	32.577308 5.166065	37.995816 5.1659	43.408056 5.164968
Aver. Bond Å: Ga-Ga;Ga-N	2.894 1.907	2.902 1.910	1.899	1.897	1.895	1.893
Average angle Ga-N-Ga/N-Ga-N	109.33 109.99	109.58 110.57	110.26 110.83	110.44 111.05	110.56 111.21	110.64 111.33
Band gap eV	5.0069	3.6191	3.5375	3.4831	3.4831	3.5375
UC Connolly surf. area Å^2	87.18	195.93	302.10	408.45	514.67	621.05
UC Pore Volume Å^3	1073.95	1634.95	2185.59	2738.86	3289.20	3836.18

Table 3. Calculated characteristics of all studied phase and most common phase of GaN-3W.

Structures	**GaN-hex12-4W331**	**GaN-hex23-4W441**	**GaN-hex34-4W551**	**GaN-hex45-4W661**	**GaN-hex56-4W771**	**GaN-hex67-4W881**
Mass Density (gcm^{-3})	6.241	5.485	4.975	4.140	3.733	3.353
Volume/at (Å^3/at)	11.139	12.675	13.937	16.792	18.623	20.731
Part.density (10^{23}cm^{-3})	106.898	166.789	226.716	286.596	346.537	406.482
Coord. Number	0.00 3.94	0.29 3.82	1.03 3.76	0.48 3.64	0.00 3.60	0.02 3.57
Cr. structure	hex	hex	hex	hex	hex	hex
Symmetry groups	P31M (C3V-2)	P31M (C3V-2)	P31M (C3V-2)	P31M (C3V-2)	P31M (C3V-2)	P31M (C3V-2)
Unit cell (atoms)	106	166	226	286	346	406
Lattice param. Å (a-c)	16.260907 5.156208	21.701118 5.158977	27.150467 4.955808	32.598009 5.227563	37.966864 5.161848	43.392612 5.162097
Aver. bond Å Ga-Ga; Ga-N	1.917	2.922 1.912	2.863 1.883	2.917 1.915	1.902	2.928 1.901
Average angle Ga-N-Ga/N-Ga-N	109.56 109.62	109.69 110.01	109.41 110.28	109.83 110.58	110.17 110.63	110.25 110.73
Band gap eV	2.3130	3.7552		3.6191	3.6191	3.6191
UC Connolly surf. area Å^2	9.58	142.67	238.13	361.67	460.89	567.21
UC Pore Volume Å^3	1177.93	1860.59	2441.78	3260.33	3905.75	4586.73

Table 4. Calculated characteristics of all studied phase and most common phase of GaN-4W.

3.2. Electronics properties

IOP Conf. Series: Journal of Physics: Conf. Series **1034** (2018) 012009 doi:10.1088/1742-6596/1034/1/012009

In order to determine would the newly designed GaN nanoporous phases, if synthesized, possess novel electronic properties, we firstly explore their band structures. Figure 6 shows the calculated of nanoporous phase's band gap vs. their hollow size. It should be noted that they all are below the GaN wurtzite value [25] (~ 4.16eV within the DFTB+ parametrization used this work). Noted that within the used scheme, DFTB+ calculation had overestimated the GaN-w band gap, which is 3.4 eV from the room temperature experimental measurements about 20%. The gaps vary from 3.1 to 4.0 eV among the nanoporous series with different wall thickness denote by different color and symbol in Fig.5. Further, our results show that all nanoporous series are still wide band gap (WBG) semiconductors with a direct gap at the Gamma point. On the other hand, we observe general trend that the gap reduced with the increasing hollow radius then approaching to a certain saturated value at the large radius for each series. The reason is following the expected behavior of the quantum confinement effect. Also there are some evidence of the surface effect observed for the smallest cage's radius structure for each series below the radius of 7A, that causes the band gap is lowering due to the in-gap position of the surface states. This effect is more visible for the thinner series, i.e. SW and DW and less visible for the thick one as 3W and 4W. More over while the SW and DW series are separated each from other, the 3W and 4W series are very close. This might shows a tendency that the thicker series would approach to a certain convergence of saturation behavior, which helps our calculation prediction for the larger wall thickness. Overall, our result shows that the nanoporous phase band gap is an important parameter which could be turned due to their dependence to hollow size and wall thickness.

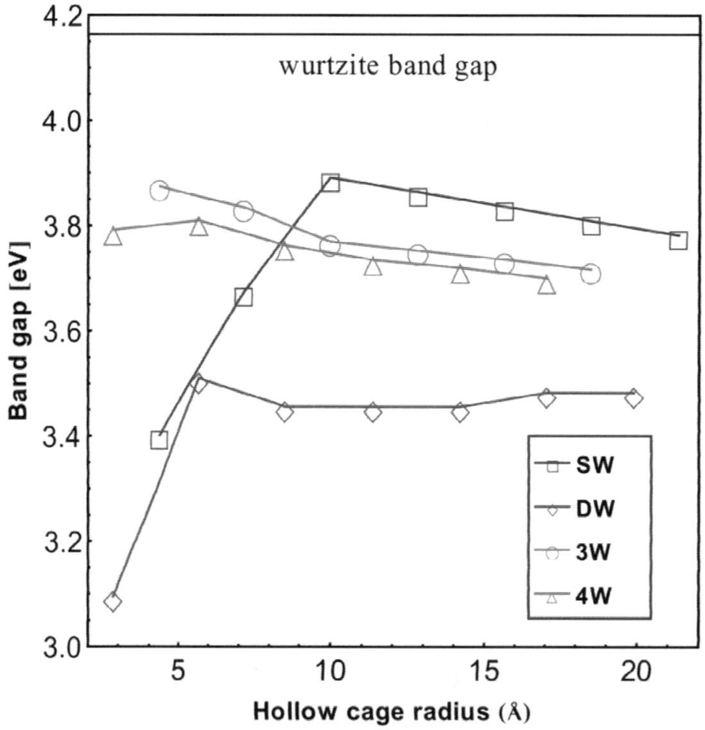

Figure 6. Band gap vs. hollow and wall thickness.

Our calculated bulk modulus (BM) are showing in Figure 7 and also listed at the Table of Characteristics, which indicates that for all series, while hollow size increases, the BM first falls rapidly then saturates at certain large hollow size. Also the thicker wall, i.e. 3W or 4W are more stable then the thinner with respect to same hollow radius as expected.

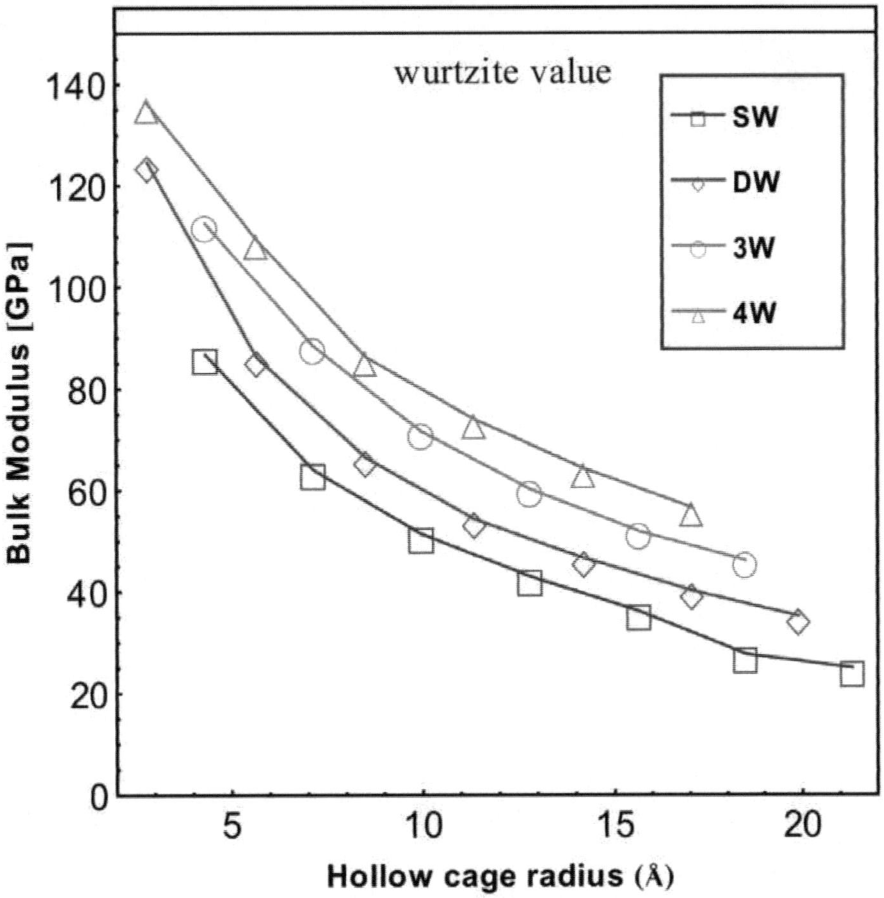

Figure 7. Bulk modulus vs. hollow cage radius.

In order to estimate the relative strength of these nanoporous phases, we have made a figure, that draw the same data of BM but with respect to the phase's *porosity* value called *hollow-to-bulk* density ratio in Figure 8. It is obviously that by increasing the hollow effective radius (denote by an arrow at the right corner) which equivalent to the increasing of phase's porosity all the BM first fall rapidly then approaching to the linearly scaling line, i.e. thin 1:1 black line in the plot, called the *strength-to-density* scaling line. Furthermore, the series with thicker wall (larger than three layers) as 3W and 4W, become very close each other with respect to the same porosity (see lines in Figure 8). Therefore, we can make an assumption that although the BM values are separated for each series with respected to the hollow size (see Figure.7). However, they are very closed and even converged (for wall over the three layers) with respect to the phase's porosity (see Figure.8). This is very important point for the applications. For example, our rule give a flexibility to choose between the desire BM value with corresponding wall thickness and how closer one can come to the of strength-to-density scaling line, which means closer to the bulk value of strength-to-density *figure of merit*.

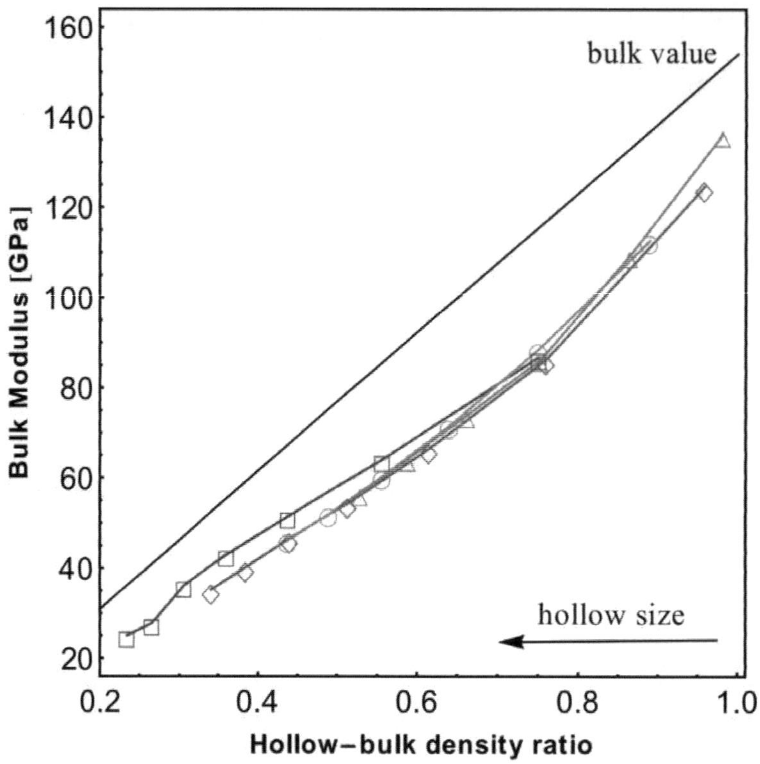

Figure 8. Bulk modulus vs. hollow-bulk density ratio.

4. Conclusion

These results show that changing the hollow size and the number of walls of the structure will change their physical and chemical properties for specific applications. This makes the nanomaterials more important. Nano-porous structures, due to their ability to interact with ions or both atoms and aggregates, make them especially useful in catalytic or membrane applications.

We believe that theoretical predictions from our simulation model can provide parameters for the design of nanoscale material models for computation. From our analyses of structure, electronics, and thermodynamic properties, it can be clearly shown that nanofunctions can actually describe nano-crystalline materials and low density. Due to the influence of the hole / wall size distribution on the properties of nanomaterials, an important parameter in practical fabrication applications is the gap- engineering.

Acknowledments

This work was supported by Vietnam National Foundation for Science and Technology Development (NAFOSTED) under grant number ***103.01-2017.24.***

References

[1] Meza L R, Das S, Greer J R 2014 *Science* **345** 1322
[2] Mezaa L R, Zelhofera A J, Clarkeb N, Mateosa A J, Greer J R 2015 *PNAS* **112** 11502-11507
[3] Bauer J, Schroer A, Schwaiger R and Kraft O 2016 *Nat. Mater.* **15** 438
[4] Tian Y, Zhang Y, Wang T, Xin H L, Li H, and Gang O 2016 *Nat. Mater.* **15** 654

[5] Xiong H, Sfeir M Y, and Gang O 2010 *Nano Lett.* **10** 4456
[6] Bauer J, Schroer A, Schwaiger R, and Kraft O 2016 *Nat. Mater.* **15** 438
[7] DFTB+ [Online http://www.dftb-plus.info/ http://www.dftb-plus.info/]
[8] Willander M 2014 *Pan Stanford*
[9] Elstner M, Porezag D, Jungnickel G, Elsner J, Frauenheim Th, Seifert G 1998 *Phys. Rev.* B **58** 7260
[10] Kohler C, Seifert G, and Frauenheim T 2005 *Chem. Phys.* **309** 23
[11] Aradi B, Hourahine B, and Frauenheim Th 2007 *J. Phys. Chem.* A **111** 5678
[12] Carrasco J, Illas F, and Bromley S T 2007 *Phys. Rev. Lett.* **99** 235502
[13] Wang Z L 2004 *J. Phys. Cond. Matt.* **16** R829
[14] Wang Z L and Song J 2006 *Science* **312** 242
[15] Moreira N H, Dolgonos G, Aradi B, da Rosa A L, Frauenheim Th 2009 *J. Chem. Theor. Comput.* **5** 605
[16] Tuoc V N 2010 *Comput. Mater. Sci.* **49** S161
[17] Perdew J P, Burke K, and Ernzerhof M 1996 *Phys. Rev. Lett.* **77** 3865
[18] Heyd J, Scuseria G E, and Ernzerhof M 2006 *J. Chem. Phys.* **124** 219906
[19] Vu Ngoc Tuoc, Tran Doan Huan, Nguyen Viet Minh, Nguyen Thi Thao 2016 *Journal of Physics: Conference Series* **726** 012022.
[20] Nguyen Thi Thao and Vu Ngoc Tuoc 2016 *VNU Journal of Science: Mathematics – Physics* **32** 1-10
[21] Vu Ngoc Tuoc, Tran Doan Huan, Nguyen Thi Thao 2017 *Physica B: Physics of Condensed Matter* http://dx.doi.org/10.1016/j.physb.2017.03.003
[22] Tuoc V N, Huan T D, and Lien L T H 2014 *IEEE Trans. Magn.* **50** 2400407
[23] Yong Y, Li X, Hao X, Cao J, and Li T 2014 *RSC Adv.* **4** 37333
[24] Vu Ngoc Tuoc, Tran Doan Huan, Nguyen Thi Thao and Le Manh Tuan 2016 *J. Appl. Phys.* **120** 142105
[25] Tuoc V N, Huan T D, and Lien L T H 2012 *Phys. Stat. Sol.* B **249** 1241
[26] Sponza L, Goniakowski J, and Noguera C 2015 *Phys. Rev.* B **91** 075126
[27] Zwijnenburg M A, Illas F, and Bromley S 2010 *Phys. Rev. Lett.* **104** 175503